Forest H. Belt's

Easi-Guide
to
SMALL
GASOLINE ENGINES

Text by Calton E. Taylor

Photography by Forest H. Belt

HOWARD W. SAMS & CO., INC.
THE BOBBS-MERRILL CO., INC.
INDIANAPOLIS · KANSAS CITY · NEW YORK

FIRST EDITION

FIRST PRINTING—1974

Copyright © 1974 by Howard W. Sams & Co., Inc., Indianapolis,
Indiana 46268. Printed in the United States of America.

International Standard Book Number: 0-672-21095-9
Library of Congress Catalog Card Number: 74-77714

Introduction

The first practical gasoline-powered engines appeared very late in the 19th century. They ushered in our Era of Easy Transportation. Quickly, gasoline engines were adapted to move horseless carriages and, soon thereafter, airplanes. The four generations since have seen fantastic improvements over those first noisy, single-piston, comparatively anemic putt-putters. Today, multi-cylinder behemoths with intricate carburetors and complicated ignition systems hurl us around in autos and planes at speeds incomprehensible to gasoline-engine pioneers.

Yet, what kind of engine do we almost invariably turn to for plain, ordinary jobs? You probably guessed it. Yes, the simple, one-cylinder gasoline engine. Look at your lawnmower, your garden tractor, snowblower, rototiller, standby electric generator, chain saw, and so on. All of them incorporate small gasoline engines that differ hardly at all from the original basic design. A raft of other small-engine-powered tools take on countless jobs that need more than one person's strength. They shorten or lighten a thousand or so other tasks we cope with every day.

You can find any number of reasons to know more about small gasoline engines. Their overwhelming number makes a good beginning. With that many of anything around, you should latch onto some idea what they're all about. You're bound to encounter some of them sooner or later. (You'd hate to be bested by a machine . . . right?)

If you own one or more small engines, an immediate and practical reason to understand them has to do with your pocketbook. Hiring someone to maintain and repair your lawnmower or other gasoline-powered tools turns out costly—supposing you can locate an expert who will find time to get to yours.

The fact is, one-cylinder engines are not all that complicated. Even with a little knowledge, you can undertake a lot of your own maintenance—perhaps even simple repairs.

This book brings you all the practical knowledge and understanding you need for minor work on your own machinery. You can see here, through explicit photographs, each part or adjustment you'll be dealing with (or one so similar to what's on your machine that you'll easily spot the differences). Clear and detailed explanations supply just about all the help you should need to apply the techniques you see in the photos. Right from the start, photos and words lay out every movement and action in a small gasoline engine, whether it's two-cycle or four-cycle.

Calton E. Taylor wrote the explanations, and prepared the engines for much of the photography. His present work as a tool-and-die designer for a major machinery company was preceded by years of maintaining and repairing engines and machinery of all sizes. We agreed some time ago on the need for a book like this on small engines—explicit, clear, and profusely illustrated. Hence this volume. This is the first book Mr. Taylor has co-authored. However, you'll soon be able to buy others we are preparing together in this *Easi-Guide* series.

We owe special thanks to Earl and Ron Christensen, the father-and-son team that operates Christensen Hardware & Repair, near Frankfort, IL. They let us use many of the engines and machinery you see pictured throughout this book, and set aside space in their excellent shop for quite a lot of our photography. Russell Murphy, owner of Small Engines, Inc., in Indianapolis, helped us set up some of the photography in his shop. We appreciate their help, and you too owe them a vote of thanks; their generosity made this book more thorough than it could otherwise have been.

One last reason for reading this book carefully—and it may turn out the most important reason of all. The energy "crisis" has brought fears that we won't have enough gasoline for our lawnmowers, chain saws, generators, etc. One key to stretching the gasoline supply falls inside the realm of this book: care and maintenance. The well kept small engine burns less gasoline per hour of operation than one that gets attention only when it won't run at all.

And gasoline, you know, is not by any means the only scarce ingredient these days. If present worldwide economic and political conditions persist, our supplies of metals and other manufacturing materials may dwindle. It behooves us to make what we now own last as long as we can; replacements may not be as easily come by as once they were.

If there is one theme underlying this book, these factors lead to it: Conserve. Learn about your gasoline-powered equipment so you can take care of it properly. It will last longer, cost you less for repairs, and economize on the gasoline it burns. We expect this book—as should others in my *Easi-Guide* series—will let you relax in the knowledge that you're treating small-engine apparatus sensibly and wisely, and saving a pocketful of money in the bargain.

FOREST H. BELT

Contents

Chapter 1

How Small Engines Work: Four-Cycle

Almost everyone knows this familiar summer scene. A lawn-mower is by far the most common vehicle using a small engine. But many other machines today use small single-cylinder engines for power. There are labor-savers: garden tractors, leaf mulchers, garbage composters, saws, hoists. Portable generators safeguard your home when electric power fails or supply current in your recreational vehicle. Speaking of fun, how about the small engine in a minibike? Not to mention minitrikes and small amphibious all-terrain vehicles.

Anyone with any machine incorporating a small engine wants long life from the engine, and efficient operation. A clean-running, properly adjusted and serviced engine starts easier, does a better job, and lasts longer. It also pays to know when to take the machine to an expert for repairs. These pages treat all this.

Mowers come in all sizes. Garden tractors have detachable mowers; they cut the lawn but are heavy enough for weeds too. The higher horsepower engine allows other jobs such as pushing snow or working a garden. Small-tractor horsepower ranges from 6 to 18 hp. Many models have electric starters.

A weed mower has the blade out in front of the engine. You can raise the blade over high weeds and let it down slowly to cut thick stalks. The rear wheels are large to provide ground clearance. A somewhat larger engine has power to cut large weeds. A self-propelled version helps even more if the weeds grow heavy or the lot is hilly.

Most riding lawn mowers do nothing but cut grass. They are faster than a walking or self-propelled mower and can be bought in many different widths and horsepower ratings to handle any size yard.

Snowblowers prevent backaches. Some can be credited with saving lives; they avoid overexerting a bad heart trying to clean up sidewalks and driveways after a heavy snowfall. Walking models are self-propelled. Other models fit on the front of garden tractors; they handle larger snow-removal jobs.

Another seasonal machine is the mulcher (at left). It chops up the leaves on a lawn in the fall; the leaves rot away through the winter and add natural fertilizer to the lawn the next spring. A series of sharp blades inside a housing work like a power beater. A large fan inside creates a vacuum to pull the leaves through the machine and blow the chopped-up pieces out the back.

The composter (at right) is a favorite of organic gardeners. It chops up leaves, grass, and whatever for the compost pile.

A portable generator produces electric power similar to the alternating current (ac) that feeds your home. A gasoline engine turns the alternator. The average portable delivers enough power to run a refrigerator, freezer, and a light or two. Construction companies use them to operate power tools where no electricity has been run to a job site.

However, small engines are not limited to work machines. Many fun vehicles have small engines for power. Minibikes are the best known. They use a standard horizontal-shaft 2 to 5 hp engine with a centrifugal clutch. You can buy a kit to build your own minibike and put on it whatever horsepower of engine you want.

Sport tricycles suit the enthusiast who wants more than two wheels under him. The drive train is the same as on a minibike. This vehicle is fun, but can't match the minibike in rough going. Nor is it as portable.

Time now to talk about how small engines work. There are two basic types. One is two-cycle; the other, four-cycle. From an operational viewpoint, the main difference is in the valve system—the two-cycle engine has none.

Briggs & Stratton is a name well known among small engine users. Other brands are similar in construction and design. We'll use it as an example of a typical four-cycle engine. The term *four-cycle* really means a four-stroke cycle, referring to the four strokes of the piston. First is intake, followed by compression, then power, and finally exhaust. These strokes are repeated as long as the engine runs.

An intake valve lets fuel into a combustion chamber. An exhaust valve lets out the burned fuel gases. Both valves operate from a camshaft that opens and closes each of them at just the right time.

An electrical spark ignites the fuel at the peak of the compression stroke. The explosively burning fuel causes the power stroke. Killing the electrical spark stops the engine.

A connecting rod connects the piston to the crankshaft. This changes the up-and-down motion of the piston to a rotary motion the engine can deliver to a load.

But this explanation is far too brief. The following pages show you each action that contributes to operation of a four-cycle small engine.

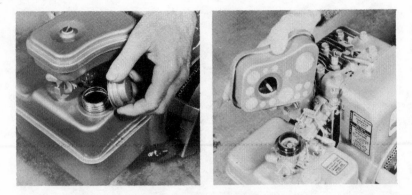

Small engines burn gasoline. A fuel tank on the machine stores a gallon or so. Buy regular gasoline; high-octane premium burns much hotter and scorches the valve edges and could burn holes in the top of the piston.

Gasoline vapor is one of the most highly explosive materials known to man. Every time the spark plug ignites the gasoline mixture in the combustion chamber, a controlled miniature explosion takes place. The powerful expansion of hot gases in the confines of the combustion chamber drives the piston downward to turn the crankshaft.

Gasoline burns well only with plenty of oxygen present. That requires a steady supply of clean air. An air filter on top of the carburetor collects dirt and impurities before the air enters the carburetor.

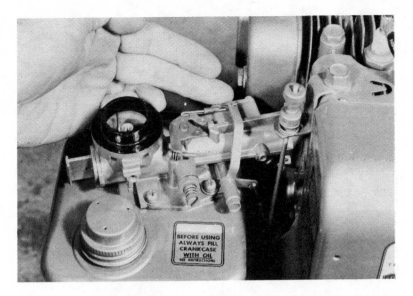

The *carburetor* mixes air and gasoline. The carburetor "feels" a vacuum, created by the piston traveling downward inside the cylinder, every time the intake valve opens. Gasoline in the carburetor is drawn through a very small hole (called a jet) into the path of the incoming air. The jet breaks the gasoline into tiny droplets that mix with the air, forming a fuel vapor.

The correct mixture ratio calls for about seven parts air to one part gasoline. This ratio must be kept constant for the engine to run smoothly. A needle-valve and seat assembly controls how much air enters the mixing barrel of the carburetor. Screw threads next to the head of the needle make the valve adjustable. Never try to adjust this fuel-air mixture unless the engine is running. The needle valve has been set properly when the engine runs smoothly at idle and accelerates evenly when given more gas.

A chamber in the carburetor, called the *bowl,* holds a small supply of fuel that has flown by gravity through a fuel line from the tank. A float-operated shutoff valve keeps the gasoline level high enough that the vacuum can draw some through the jet, but prevents overflow.

The throttle handle on a mower connects to the carburetor by a movable cable. The cable pivots a round disc inside the carburetor. The disc is called the *butterfly* valve. The angle of the butterfly valve determines how much air can enter the carburetor.

"Opening" the throttle increases the air flow, which in turn carries more fuel (as vapor) to the combustion chamber inside the engine. You affect how fast the motor runs by increasing or decreasing the amount of air-fuel mixture allowed through the carburetor.

A *governor* rod also connects to the butterfly valve. The governor keeps the engine from running too fast whenever the engine work load is small. When you mow into heavy grass or weeds, additional gasoline mixture is added automatically to increase power. When the load decreases again, the governor cuts down the fuel-air mixture going to the combustion chamber, slowing the engine to a safe operating speed so nothing inside the engine breaks apart from overspeed.

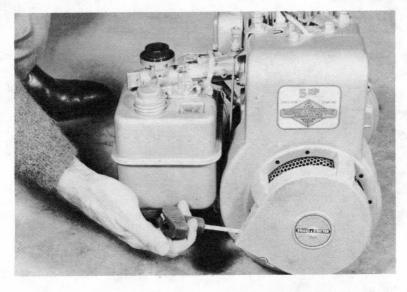

A small gasoline engine can't run unless some outside impetus starts its piston and crankshaft moving. The most common starting mechanism for small engines is a spring-rewound rope. The rope is coiled around a pulley inside the starter housing. When you pull the rope, the pulley engages a flywheel and turns it. The flywheel bolts to one end of the crankshaft. A key keeps it from spinning helplessly on the tapered end of the crankshaft. Hence, when the flywheel is turned, the turning crankshaft moves the piston up and down inside the cylinder. That draws the fuel mixture into the cylinder as already described. Add an electrical spark at the proper time, and the engine starts running on its own. A spring recoils the rope around the pulley, which disengages from the flywheel when the engine starts.

Some small engines have an electric starter. Operation is similar, with a battery-powered starter motor turning the flywheel and crankshaft. These usually have a rope-starter pulley too, in case the battery goes dead or the electric starter stops working.

Another type of starter mechanism is the *impulse* starter. A folding crank-handle on top of a starter housing winds up a large spring similar to a clock mainspring. The spring attaches to a pulley resembling that in the rope starter. A movable stop on the side of the housing holds the pulley while you wind up the spring. Trip the stop and the spring unwinds, turning the flywheel. The crankshaft turns and the engine can start running.

Here's what happens inside the engine as the flywheel and crankshaft turn. The first stroke of each engine cycle is the *intake* stroke. The piston travels downward. The intake valve opens up, allowing the air-and-fuel mixture to be sucked into the combustion chamber. (The piston traveling downward creates a vacuum which draws the fuel mixture into the combustion chamber with force.) Camshaft timing must be precise, so the intake valve opens just as the piston starts downward. Once the piston reaches the bottom of its stroke, the intake valve closes, trapping the fuel vapor above the piston inside the cylinder.

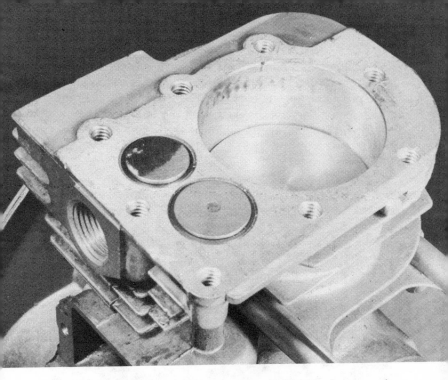

The crankshaft starts the piston upward after the intake valve closes, initiating the *compression* stroke. The fuel-air mixture is trapped between the piston and the top of the cylinder. The upward-traveling piston compresses the vapor tightly. At slightly past the top of the stroke, the compressed fuel mixture is ignited. The more tightly the mixture has been compressed, the more powerful the explosion.

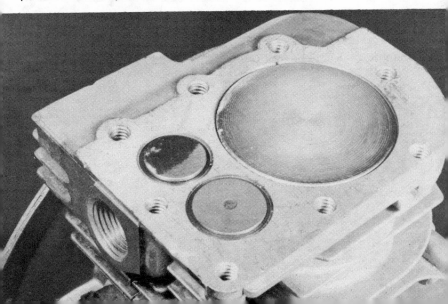

Let's pause for a moment in "mid-cycle" to consider how the fuel vapor is ignited. At the end of the compression stroke, the spark plug carries an electrical spark into the combustion chamber. Timing must be such that the spark arrives instantly after the piston passes top-dead-center. If the fuel mixture were ignited at exactly top-dead-center, the explosion would exert piston pressure straight downward against the crankshaft instead of turning the crankshaft. If the spark fires the fuel very much past top-dead-center, some of the compression is lost and the explosion develops less power.

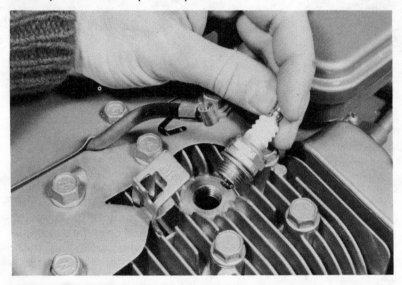

The spark plug screws into a finned plate on top of the cylinder. The plate is called the *cylinder head*. Inside the head, beneath the spark plug, is a cavity; that's the *combustion chamber*. This is where the fuel mixture is compressed by the piston.

A *spark plug* has three essential parts. The *electrode,* a small rod running through the middle, carries an electric current from the top of the spark plug down into the combustion chamber. When the current reaches the end of the electrode, it jumps across the gap between the electrode end and the grounding tip. That creates the hot spark that ignites the fuel vapor.

The *grounding tip* is part of the threaded end of the spark plug. The gap between it and the electrode must be just right. If it's too wide, the current can't jump soon enough and the spark occurs too late. Delayed firing lets compression diminish before the fuel mixture is ignited. A too-narrow electrode gap creates a spark that is not hot enough to ignite all the fuel mixture in the combustion chamber. Either fault results in lack of power from the engine and a waste of gasoline.

The third part of a spark plug is the *insulator.* That's a piece of porcelain around the electrode to keep the electric current from jumping to ground before it reaches the combustion chamber. Porcelain withstands the high temperatures produced by the burning fuel mixture inside the combustion chamber.

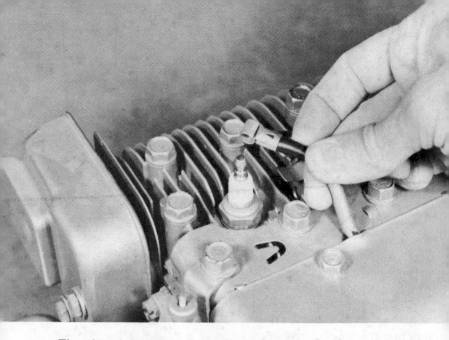

The electric current comes to the spark plug through a heavy wire. It's called the *ignition wire* or *spark plug wire*. You can see it attached to the top of the plug.

The other end of the wire goes to the *ignition coil*. The electricity developed here has very high voltage, so insulation on the ignition wire must be thick enough to prevent electrical current from jumping (making a spark) to the engine housing before reaching the spark plug. Electricians call this kind of wire high-tension wire because its insulation can withstand such high electric voltage.

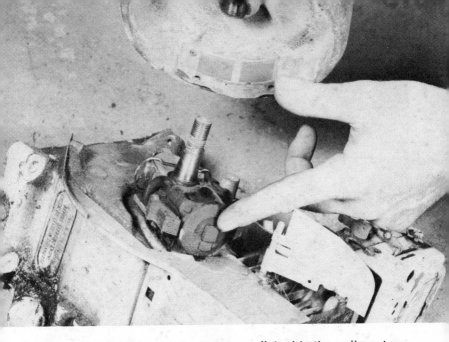

The spark-plug wire originates at a *coil.* Inside the coil are two separate windings of fine copper wire. The coil forms half of a device known as a *magneto.* A group of permanent magnets set in one side of the engine flywheel forms the other half. This arrangement of magnets and coil can magnetically generate the electric current to make the ignition spark. The engine thus doesn't need a battery to operate the engine.

Here's how the magneto works. Magnetic fields exist around the permanent magnets. As the flywheel is turned, the magnets pass by the coil, which fits adjacent to the flywheel. The coil intercepts the magnetic field as the permanent magnets move past and the magnetic forces induce an electrical surge through the wires in the coil.

One winding of the coil carries the electrical surge to a set of breaker points. They short out this winding most of the time, and the effect of the induced electric surge is almost nil. However, at just the right instant after the surge begins, the breaker points open. This coincides with the instant the spark is needed in the cylinder. Without that quenching short, the electric surge transfers to the secondary winding of the coil. The secondary winding steps up the voltage, making it strong enough to jump the gap at the bottom of the spark plug. A proper high-voltage surge creates a spark that is intense enough to burn the fuel mixture cleanly in the combustion chamber.

And that brings us back to the third stroke in a four-stroke cycle: the *power* stroke. When the fuel mixture ignites, the resultant expansion of gases (a controlled explosion) drives the piston downward. The power stroke delivers enough force to the crankshaft and momentum to the flywheel to keep the engine going until the next power stroke. The stroke must also deliver enough power to do the work the engine is responsible for.

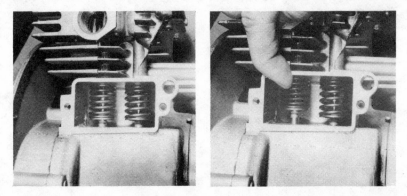

Intake and exhaust valves are both held closed by heavy springs. A beveled washer with a hardened pin behind it holds each spring in place. The pin that holds the washer fits tightly in a hole in the valve stem.

A *camshaft,* turned by the crankshaft, opens the valves at the right time. Gears between crankshaft and camshaft time the opening and closing.

The camshaft pushes a rod-like stem called the *valve lifter.* Each valve has its own. This piece of hardened and ground steel has one end resting against a lobe on the camshaft and the other against the valve stem. As the camshaft spins, the lifter and valve "ride" on the lobe. Where the lobe is high, the lifter and valve move upward, raising the valve up off its seat and opening the hole. If it's the intake valve, fuel mixture rushes into the cylinder. If it's the exhaust valve, burnt gases are expelled through the hole (see next page).

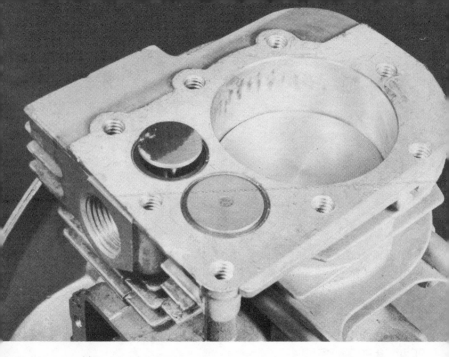

The *exhaust* stroke is the last of the four. After the power stroke drives the piston down, flywheel momentum turns the crankshaft and starts the piston back upward. The exhaust valve opens. The traveling piston pushes out all of the burnt gases and fumes. This clears out the cylinder, which is then ready for the next charge of fuel-air mixture to be drawn into the cylinder. The exhaust gases leave through a hole in the side of the engine block. A noise muffler (facing page) fits on the outside of the exhaust port.

That describes all four stokes of this type of engine. After the piston reaches the peak of the exhaust stroke, the exhaust valve closes. The intake valve then opens and the piston starts downward again on the next intake stroke. This entire sequence repeats as long as the engine runs.

The speed of an engine can be raised by increasing the amount of mixture that reaches the cylinder. Allowing more air to enter the carburetor barrel draws more fuel into the barrel too. More fuel-air mixture in the cylinder makes a more powerful explosion during combustion. The piston moves faster which creates more vacuum, which draws more fuel mixture, and so on. Speed of the engine depends on the balance between vacuum and the air-fuel mixture.

Any explosion creates a lot of noise. Since a small engine generates around 1500 explosions every minute, you'd expect some noise—and loud. The noise reaches the outside of the engine when the exhaust valve opens. That noise could damage ears, and must be reduced to an acceptable level.

A muffler has been attached to the engine directly outside where the exhaust leaves the cylinder head. All the gases and the noise from the engine must pass through this muffler before reaching the outside air. Noise is damped considerably.

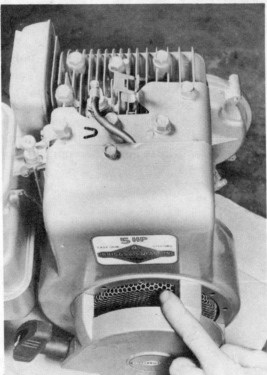

The confined burning of the fuel mixture inside the cylinder creates intense heat. This heat must be transferred away from the combustion area very quickly. Otherwise, the engine would soon get overly hot. The piston would expand, tighten up in the cylinder, and stick to the cylinder walls. That would ruin the engine.

The outside of the cylinder and the head have fins cast on them to help transfer the heat efficiently to the outside air (see facing page, top). The walls of the cylinder are just thick enough to withstand safely the pressures of explosions inside the cylinder. The relatively thin walls transfer the heat quickly to the outside fins. The metal fins expose more than twice as much surface to the air as an equivalent flat area could. They exhibit twice the cooling effect the cylinder walls alone might.

To provide even more cooling, the flywheel (below) has a row of fins cast around the top of its outer edge. This row of fins works like a fan, drawing cool air from outside the engine past the cooling fins on the cylinder head. The flywheel fins are directly under the starter housing (opposite page, bottom), which acts like a funnel to direct the incoming air to where it can best cool the engine.

Small gasoline engines sold separately (not as part of a tool or machine) are often categorized by the drive characteristics. For example, a *horizontal-shaft* engine has the crankshaft in a horizontal plane and the cylinder vertical. This type engine fits where power must transfer from the crankshaft directly to the wheels of a vehicle.

A common application of the horizontal-shaft arrangement is in minibikes. A centrifugal clutch on the end of the crankshaft couples crankshaft rotary power to a roller chain and thence to the rear wheel of the minibike. A V-belt drive permits the use of a third pulley to let the operator tighten the belt manually, eliminating the cost of a clutch.

Another typical use of the horizontal-shaft engine is in garden tractors. Most of these vehicles have a V-belt drive from the engine to a transmission; a third pulley may serve as clutch. The transmission incorporates two or three different gear ratios to provide appropriate ground speeds and adequate torque (mechanical muscle) to do the work required.

Some horizontal-shaft engines come with a small gearbox mounted on the engine. This setup converts the crankshaft rpm's to a slower speed. As a transmission does, this gearbox delivers more output torque, so you can get more work leverage from a smaller-horsepower engine.

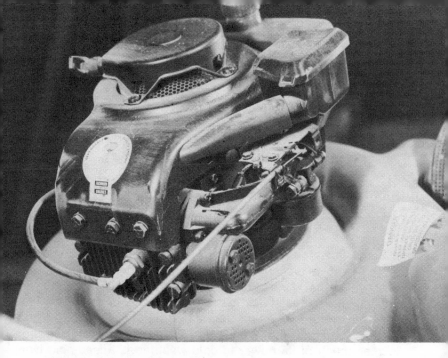

In *vertical-shaft* engines, the crankshaft delivers its power in a vertical plane. The cylinder operates in a horizontal position. This is the type of engine used on rotary-blade lawn mowers. The mower blade attaches directly to the end of the crankshaft. On self-propelled models, a V-belt pulley on the end of the crankshaft drives a right-angle gear-and-axle system to turn the wheels of the mower.

Vertical-shaft engines also power articulated (jointed in the middle) riding lawn mowers. The blade and a V-belt pulley mount directly on the crankshaft of the engine. The V-belt runs to a single-speed reversible transmission. The transmission allows for forward and backward movement of the mower. A roller chain drive carries power from the transmission to the rear wheels.

Some late-model walking mowers have an electric starting system with a *nickel-cadmium* (NiCad) battery to supply the energy. This new kind of battery delivers more electricity than an equivalent small lead-acid battery. The nickel-cadmium battery has a much slower shelf-discharge rate than a lead-acid type, and can be recharged with a trickle charger. A NiCad battery does not lose as much of its power in cold weather as a lead-acid type does.

Too, the lead-acid battery is messy. The NiCad comes sealed, and never needs electrolyte or water added. The plates in a NiCad battery are made of two different metals rather than one as in a lead-acid battery. The metals, obviously, are nickel and cadmium. The liquid electrolyte is potassium hydroxide.

A nickel-cadmium battery should last 7 to 10 years in normal use. You should be careful about overcharging a NiCad battery, since that reduces its life. When you buy this kind of mower, you usually get a special charger that eliminates the danger of overcharging.

Most garden tractors have an electric starting system like a car has. The system uses a lead-acid battery (next page), like a car battery only smaller. The starter housing also has a generator built in, to keep the battery charged. The battery and starter merely turn the engine; a magneto provides the electric spark to run the engine. (In a car, the battery operates the ignition system as well as the starter.)

The electricity from the battery must go through a *solenoid* to reach the starter. The solenoid is a heavy electromagnetic switch that also engages and disengages the starter gear from the flywheel gear. The starter itself is an electric motor with enough power to turn the engine until it starts running on its own.

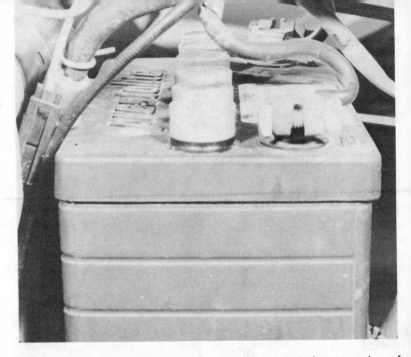

The lead-acid battery used for starting power has a series of lead plates inside a leakproof plastic case. The electrolyte, the liquid that surrounds the lead plates, is diluted sulfuric acid. The electrolyte interacts with the lead plates and creates electricity.

In the process, bits of the plates inevitably dissolve and form a sediment in the bottom of the case. If the sediment builds up enough to touch two of the plates, the battery shorts out and can no longer hold a charge.

One remedy for this condition, you can try yourself. Just add a half-teaspoon of baking soda to each filler hole at the top of the battery. Leave the sodium bicarbonate in the battery overnight. In the morning, turn the battery upside-down and drain out all the old acid. Tap the bottom of the case lightly with a piece of wood or something like that. This breaks loose the sediment and causes it to fall out of the battery. Then refill with new sulfuric acid, which can be purchased from a battery dealer, and recharge the battery.

You can prevent sediment from building up. Keep the acid level in the battery up to the mark. Also, never let a battery discharge entirely. Finally, when you're recharging a battery, always use a trickle charger; the slower charging rate forms less sediment.

The *solenoid* is that heavy duty switch mentioned earlier. When you turn the starter key to the "Start" position, electricity runs to the solenoid through the small wires. An electromagnet inside pulls two heavy contacts together. This couples electricity from the heavy cable that leads from the battery to another heavy cable that goes to the starter. The starter motor turns the engine flywheel and crankshaft.

The heavy cables carry high-amperage electrical current. That's why they're extra heavy—to reduce the heat that would be generated in small wires by so much electricity traveling through them. If the cable were to heat up, much of the electricity would be lost before reaching the starter. It would turn the engine too slowly for starting. (That happens anyway if the heavy-cable connections get loose and make poor contact at the battery, at the solenoid, or at the starter.)

A small engine ordinarily is stopped by eliminating the electric surge that produces a spark in the spark plug. A movable steel tab mounted next to the spark plug makes the simplest ''kill'' mechanism. When you want to shut off the engine, you push this tab against the top of the spark plug. Electricity from the magneto is grounded to the engine housing before it can travel through the spark plug into the combustion chamber. Without any spark to ignite the fuel mixture, the engine cannot continue to run. It stops when the momentum of the crankshaft and flywheel is used up.

You'll see more elaborate kill systems further along in this book.

Chapter 2

How Small Engines Work: Two-Cycle

On the outside, two-cycle engines resemble four-cycle types. The term *two-cycle* means that a mere two strokes of the piston complete each power cycle. Two-cycle engines seem to last longer than four-cycle models.

Inside, you find the main difference between two-cycle and four-cycle engines: their construction. A two-cycle engine has no opening/closing valves to let the fuel mixture in and the burnt fuel gases out. In their stead you find holes or ports cast in the sides of the cylinder. The fuel mixture enters through one set of ports and the exhaust exits through others.

A two-cycle engine has no oil reservoir around the crankshaft. Oil for lubrication is mixed with the gasoline. The proper ratio: thirty-two parts of gasoline to one part of oil. That's 1 oz of oil for each quart of gasoline, or 4 oz per gallon. A specially refined oil is sold for use in two-cycle engines.

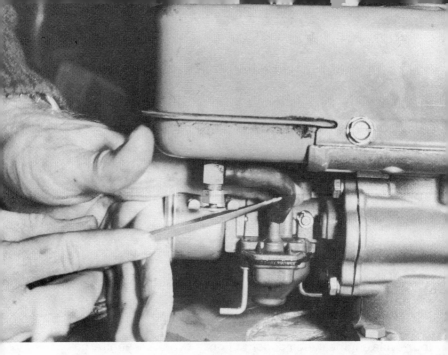

The fuel system is almost the same as on four-cycle engines. A gallon or half-gallon tank holds the fuel supply. The gasoline travels through a short fuel line from the tank to the carburetor.

The jets in a two-cycle carburetor are comparatively large, since oil must pass through along with the gasoline. The size of the air intake is about the same as on a four-cycle carburetor, but it does take more air than with an oilless mixture.

The fuel-air-oil mixture proceeds down the barrel of the carburetor, and into the combustion chamber through intake ports in the cylinder wall.

The gasoline-and-oil fuel mixes with air in the carburetor the same as for a four-cycle engine. The fuel squirts into the intake air stream via jets that break it into tiny droplets. It and the air form a vapor that will explode in the combustion chamber instead of just burning mildly. One reason the ratio of air to fuel is high (more air) is to force the mixture all around the crankshaft, the piston, and the cylinder walls. That's necessary for lubrication.

An electric current surge to create the spark that ignites the fuel mixture in the cylinder comes from a magneto to a spark plug, just as in the four-cycle engine. But in the timing of that surge, two-cycle operation introduces a major difference: A spark is delivered to the combustion chamber *every time* the piston passes top-dead-center instead of every other time.

This requirement spaces the permanent magnets around the flywheel at intervals that are different than for four-cycle operation. There is also no need for a camshaft to open and close valves in the right time relationship to the spark. The timing in a two-cycle engine involves only crankshaft and flywheel (because the magneto is part of the flywheel). To keep those two correctly related, the flywheel is locked on the crankshaft by a key that prevents the flywheel slipping during operation.

Magneto operation is explained on pages 21 and 95, in detail. Permanent magnets cast into the outside diameter of the flywheel pass by the coil. Disruption of the magnetic field around the magnets induces a pulse of electricity in the first (primary) winding of the coil. Breaker points keep the electricity quenched until they open, just at the right instant. This times the surge in the coil's secondary winding so the high-voltage surge is delivered to the spark plug at exactly the moment to ignite the fuel mixture.

For some two-cycle lawnmower engines, the head and cylinder sleeve are cast as one piece. This design removes the possibility of a blown-out head gasket and eliminates the ensuing maintenance cost of having the gasket replaced.

Intake and exhaust ports are cast into both sides of the cylinder walls. Exhaust ports extend through the cylinder wall to outside the engine. The intake ports are more complex. Grooves outside the cylinder wall run parallel to the inside bore. Intake holes (ports) then give the fuel access to the inside of the cylinder.

Travel of the piston up and down the cylinder alternately exposes and covers the intake and exhaust ports, each at different times during the two-stroke cycle of operation.

A connecting rod couples the piston to the crankshaft assembly. The crankshaft transfers the power of the exploding fuel mixture to the mower blade or whatever does the work of the particular machine.

Fuel vapor passes all around the crankshaft and rod on its way from the carburetor to the intake ports. The oil mixed in the fuel offers the only lubrication for the crankshaft and rod. Too much oil makes combustion poor and the engine can't deliver the expected power. Yet, if the mixture doesn't contain enough oil, the crankshaft and rod starve for lubricant and wear out too fast. Forget the oil, and the engine will starve and lock up in a few minutes of running. You'll ruin the engine completely.

The crankcase is made with grooves along the inside surfaces (opposite page) to direct the fuel mixture to the intake ports. The construction also circulates the fuel mixture around the crankshaft and rod assembly to permit proper lubrication.

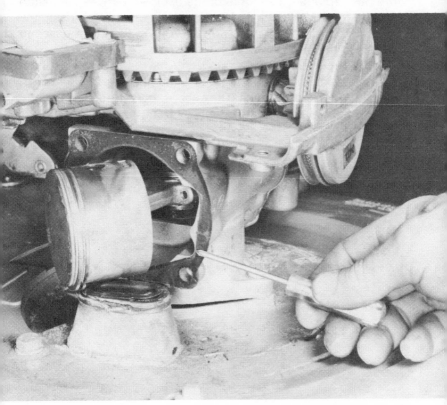

The downward stroke of the piston during the two-stroke cycle has two functions. Old exhaust gases are expelled and new fuel is taken in—both during one downstroke. The exhaust half of the downward travel of the piston begins shortly after the fuel mixture has exploded. The force of the explosion drives the piston downward (power stroke). After a certain amount of travel, the piston exposes the exhaust ports, allowing the burnt gases to escape to the outside through a muffler.

Some two-cycle lawnmowers don't have a separate muffler. Instead, the muffler is built right into the housing of the mower, directly beneath the engine. The exhaust comes out underneath, and helps scatter the cut grass away from the mower.

The intake portion of the stroke is the lower half. The intake ports are placed lower on the cylinder wall than the exhaust ports.

When the intake ports are exposed, new fuel mixture is drawn into the combustion chamber by the partial vacuum left by the exhaust gases leaving the cylinder. This vacuum also draws more fuel mixture from the carburetor barrel into the crankcase where it will lubricate the rod and crankshaft.

The effects of the vacuum reach all the way through the carburetor. It causes the strong intake of air through the air cleaner that covers the carburetor throat. And the rush of air is what vaporizes the fuel mixture in the carburetor.

The kill (shutoff) system on most two-cycle engines consists mainly of a wire run from the magneto to a metal tab. Touching the metal tab with any other metal that is grounded to the mower housing shorts out the electric surge before it even develops into a high-voltage surge. Then no spark enters the combustion chamber.

One way is with a knob that has a metal tab on it. When the knob is turned so that its metal tab touches the other one (from the magneto), the electricity is shorted to the engine housing.

On one engine, the same knob is situated above the chamber of the carburetor. Pushing the knob down squirts a little raw fuel into the carburetor air stream. This puts a richer fuel/air mixture (more fuel) into the combustion chamber. This is called *priming,* and is very helpful when you're trying to start a cold engine —especially on a cold morning.

Chapter 3

When They're Hard to Start

There are specific steps to starting a small gasoline engine. Most complaints of hard-to-start stem from poor starting procedure—not from defects in the engine.

Perhaps the first thing for you to learn is the right way to pull a starter rope. Wrong technique makes the engine harder to start because you can't spin the engine fast enough.

Never try to "hold" the engine with one hand (below) and pull the rope with the other. You would need shoulders like a weightlifter to spin the engine fast enough to start. Instead, put one foot on the mower housing in front of the engine. Place the other well away from the mower. Lean over the mower slightly and then lean back as you pull the rope with a sharp jerk. This stance and movement lends the power of your whole body to the pull. You remain balanced on your feet, and the mower can't move around. (More of this—with photos—on pages 50 and 51.)

Flooding is another typical reason an engine may be hard to start. The condition develops from too much fuel reaching the combustion chamber and not being burned.

When an engine is cold, the fuel mixture turns rapidly from vapor back to liquid gasoline. Then the piston can't compress it enough to make ignition possible. Flooding usually is caused by choking the cold engine too much. The choke is simply a butterfly valve in the carburetor throat that cuts down how much air enters the carburetor. A little choking, as you'll see, makes the fuel richer on a cold morning and aids starting. Too much makes an overrich fuel mixture, and prevents combustion.

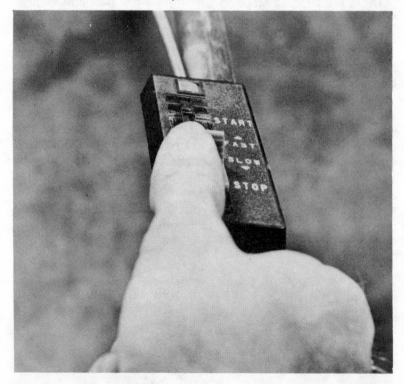

Cold weather (opposite) makes any engine hard to start. If you plan to use the machine in winter, change the oil (four-cycle) to a lighter weight. Always check your owner's manual for the proper weight of crankcase oil (and for two-cycle fuel mixing) for both winter and summer. Remember also that colder temperatures let the engine flood easier. Use the choke sparingly.

A NORMAL STARTING PROCEDURE is easy to learn. The next few pages show the right way.

Before you try to start any new engine, read the owner's manual. It helps you understand the peculiarities of a particular engine. It shows all of the controls on the machine and describes exactly how to use them. The manual also relates the service information you need for the engine.

It's no joke about instruction manuals, even though most people ignore them or treat them casually. They can save you a lot of aggravation, wasted time, and needless expense.

1. THE FIRST STEP in attempting to start any engine has to be: Check the fuel level. Don't trust memory. Remove the cap and stick your finger into the fuel tank. If you can see into the tank, okay, but that doesn't always tell you how much fuel there is. An even better procedure is to go ahead and fill the tank before each use.

2. NEXT, if the engine is on a lawnmower, lift one side or the front of the mower housing and clear all obstructions from beneath the mower. Never try to start a mower that's sitting in a driveway. Small rocks often get thrown from under a mower when you start it. They damage the blade and could easily hit someone (you) and do serious injury.

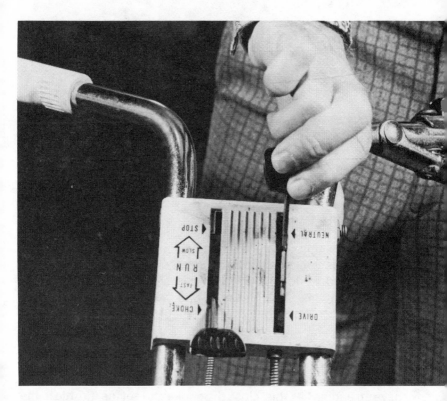

3. DISENGAGE the drive unit of any self-propelled machine before you try to start it. This reduces considerably the amount of pull needed to start the engine, and keeps the unit stationary while you tug the starter. Furthermore, you don't want the machine to take off on its own—perhaps over your own toes.

When an engine has been shut off for a half-hour or more, or the first time you start it for the day, consider it a cold engine. Your procedure for STARTING A COLD ENGINE differs from when the engine is hot. This page and the three that follow show and describe steps that apply particularly to starting a cold engine.

4. Having taken preliminary steps and safety measures 1 through 3, set the throttle lever to the Choke or Start position. On some engines, the choke lever is separate from the throttle. If you have an engine like this, close the choke and set the throttle to the Idle position.

5. Then set your hands and feet in the proper position (photo) to pull the starter rope. Be sure you are balanced before you attempt to start the engine. Keep feet away from the edges of a mower, except for one on top for bracing.

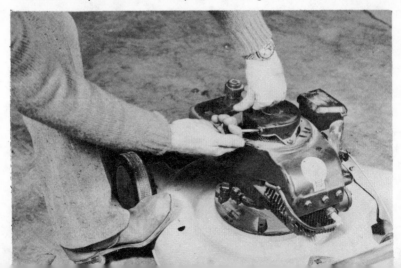

6. The direction you PULL THE ROPE affects how easily you can start a small engine. Never pull the rope upward (top photo). This reduces the power from your arms, strains the rope, and throws you off-balance.

The proper way: Pull the rope straight outward from the starter, or as as nearly straight out as possible. If the starter doesn't happen to be mounted flat on top of the engine, always pull the rope so it stays in line with the pulley the rope winds around. Don't "bend" the rope so it chafes against the edge of its guide hole. That wears it out fast.

Use a fast, firm pulling action. The engine needs to turn fairly fast in order to take off on its own. An engine in good shape should start, even when cold, in four of five pulls.

6A. If your engine has an impulse starter, turn the crank till the spring is as tight as possible. These starters have a built-in stop; you can't hurt the spring by winding too tight. If the spring isn't wound tight enough, an impulse starter won't turn the engine fast enough to start it.

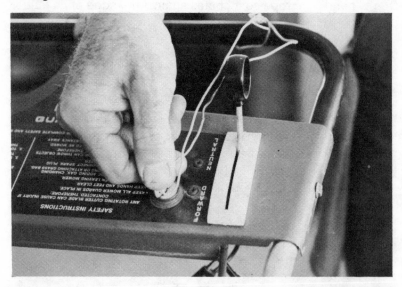

6B. With an electric starter, turn the engine only four or five times for each try. Release the starter key, wait 5 seconds, and try the starter again. Repeat this cycle several times till the engine starts. Never make the starter keep turning the engine over and over. This only runs down the battery faster and won't start the engine any quicker.

7. When the engine starts running on its own, quickly move the throttle lever to the Run position. If the engine quits, repeat the procedure and move the throttle lever to Run more gradually next time the engine starts.

The engine may even have to run with the choke partially closed for a few minutes, until the engine starts to warm up. This is especially likely in wintertime. An older engine usually takes a bit more choking than a new or freshly overhauled one.

A HOT ENGINE takes some small changes in starting procedure. How do you know what to consider a hot engine? If it's been running, and was shut down within the last half-hour, it probably still has enough engine heat inside to start without the choking and extra trouble a cold engine needs. Fuel vaporizes and ignites easily in a warm engine, and starting is easy.

Touch the head of the engine near the spark plug with a wet finger. If it sizzles, or if you can't lay your hand tightly against the head without discomfort, the engine is hot. Treat it as such. Use the procedures on the following two pages.

8. A hot engine should start with one or two pulls of the rope and with the throttle just above Idle position. If it doesn't, you may have to choke the engine slightly. The handiest way to accomplish this in just the right amount is to pull the starter rope through slowly just one time with the choke closed. Then reopen the choke and set the throttle to the Run position. Now pull the rope quickly and firmly (pages 50 and 51) four or five times.

8A. With an electric starter, close the choke and run the starter for only one turn of the engine. Then reopen the choke, set the throttle to Run position, and crank the engine four or five times with the starter.

8B. On engines with an impulse starter, wind up the starter spring only halfway. Close the choke. Punch the release and the engine rotates a few turns. Reopen the choke and reset the throttle to Run position. Then wind the spring tight and crank the engine as usual.

9. After choking the engine slightly as described on the preceding page, reset the throttle to Run position. Pull the starter rope repeatedly as rapidly as you can. The engine should take off and run. You should only need to work at these last two steps three or four times to get the engine running. If it takes more, it's time for procedures that are more complicated.

THE HARD-START PROCEDURE that follows on the next five pages—steps 10 through 14—lets you start just about any engine under almost any conditions. The exception is an engine that needs servicing. But assuming nothing is wrong with your engine, these five steps, followed in sequence, can do the job for you.

You can begin by repeating the steps for starting a cold engine (pages 50–52). Reread these steps so you don't forget any and can be sure you're following them correctly.

Use the rope starter, even if your engine has an electric starter. There's almost always a pulley on the flywheel that you can wind a pull-rope onto. The extra turning speed you can give the engine delivers a hotter spark and makes starting easier.

10. The failure to start might indicate the fuel tank outlet is clogged. No gasoline can reach the carburetor. Tip the machine both ways. This shakes sediment loose and the sloshing should wash it away from the outlet. It may settle again, but your shaking will let enough fuel through to start the engine. You'll have an idea of the trouble and can flush the tank later.

If you'll always strain the gasoline you put in the tank, you can avoid this trouble. Also, a machine should never be stored with fuel in the tank. Additives can gum up the fuel lines. (See pages 136 and 137 for more about this.)

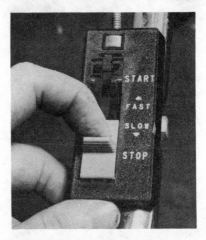

11. At this point, close the throttle completely and kill the spark. With the throttle closed, no fuel reaches the cylinder. As you crank the engine, some air enters, and that helps evaporate any oversupply of raw fuel that may be in the cylinder. Pull the rope through five or six times with everything shut off. This should clear out any possibility of flooding that might keep the engine from starting.

12. Now set the throttle to its Idle position and pull the rope to its full length rapidly three or four times. If the engine tries to start, pull the rope through again a few times. This should start the engine. If not, go on to step 13.

13. Set the throttle at the Run position. Now pull the rope rapidly three or four times. Steps 11 and 12 relieved any flooded condition that may have existed. You may now have need for a little extra fuel to reach the cylinder. This step, pulling the starter rope with the throttle set to Run, feeds in enough fuel for average-condition starting, but introduces little likelihood of flooding. Pull the rope firmly to its full length several times.

14. If the engine still hasn't started, go back to page 50 and repeat the entire procedure for starting a cold engine. Begin with step 4. Set the throttle to Choke or Start position and give the whole procedure one more run-through. Pull the rope as rapidly as possible, several times. If you reach this step, number 14, again without the engine starting, turn the page.

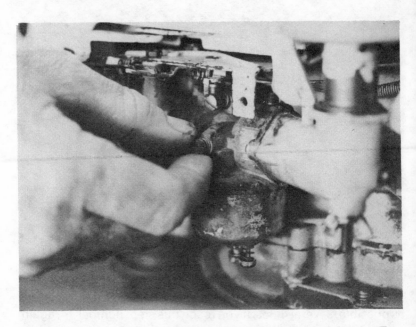

DO NOT DO THIS unless you are a qualified machanic. Too many people fiddle with the carburetor screws when the engine won't start. Mostly, that's because the screws are easy to reach. Never touch these screws unless you know how to give the engine a tuneup. Even then, you don't adjust them until the engine has started.

The fact is, if the engine ran okay the last time you used it, nothing will have changed the setting of those carburetor screws in the interim. So leave them alone. Look somewhere else for the trouble.

You'll need MORE HARD-START POINTERS if the choke and throttle controls are the type you find with a minibike engine. For one thing, you'll have to hold the throttle in position with one hand and pull the rope with the other. The choke is separate; you merely set it open, or halfway, or closed.

Step 1, naturally, is: Check the fuel level in the tank. You can see the gasoline with the cap off the tank, if the tank is full or close to it. Otherwise, use your finger to check the fuel level. It's always best to fill the tank before you start the engine. That makes sure you have plenty of gas, and you don't spill gasoline accidentally on a hot engine—potentially a dangerous thing.

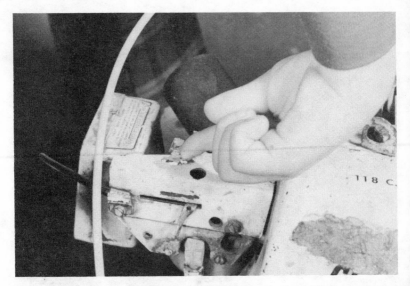

Step 2. The choke should be open on your first attempt to start the engine. This keeps you from flooding the engine on the first pull-through.

Step 3. Leave the throttle closed. Pull the rope to its full length slowly three or four times. This starts the fuel flowing into the carburetor, getting it ready for the incoming air to pick up and take to the cylinder.

Step 4. The throttle of a minibike engine is at Idle when the twistgrip is in the closed position (top rolled away from you). When you go to start the engine, you'll have to hold the twist-grip in one hand all the time, so you can open up the throttle when the engine starts.

Now, with the throttle held closed and the choke still open, pull the rope five or six times rapidly. This should apply enough fuel mixture to start the engine.

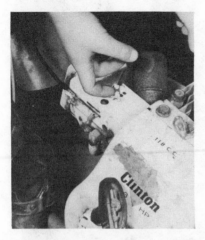

Step 5. The engine may not start without some choking. Next, set the choke lever to the halfway position. That aligns the choke butterfly in the carburetor so that less air gets in. Result: a richer fuel-air mixture for the combustion chamber in the engine.

Turn the twistgrip just enough to crack the throttle open slightly. This sets the carburetor-barrel butterfly valve so some fuel goes to the cylinder when you try starting the engine. Now pull the rope rapidly five or six times.

Step 6. If that didn't start the engine, move the choke bar to the fully closed setting. Hold the throttle the same as in step 5 and pull the rope to its full length five or six times more

in rapid succession. This step provides all of the choking the engine can possibly need to start it. There's even an outside chance that the engine has flooded—gone past the "happy medium" of just the right mixture.

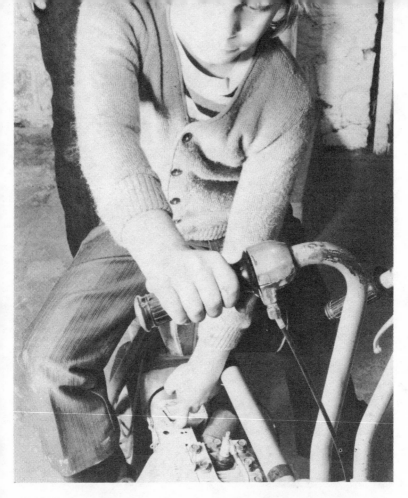

Step 7. As a last resort, reset the choke to the Open position. Let the engine sit a few minutes without bothering it. Then open the throttle about one-third to one-half of the twistgrip travel. Holding it there, pull the rope rapidly until the engine starts. As you're pulling the rope, let the position of the throttle roll gradually back toward the Idle position. This reduces the amount of fuel mixture being drawn into the cylinder and helps prevent flooding the engine.

If you don't have the machine running by now, it needs something more than a hard-starting procedure. The engine probably has a fault. You can read Chapters 4, 5, and 6 to find out what else you can do. The next two chapters (4 and 5) detail some of the proper ways to test an engine to find out what's wrong with it. Chapter 6 shows you how to fix minor ailments that prevent an engine from starting, but without going into extensive repairs. Those are for an expert, not for you (unless you have the training).

Chapter 4

Tracking Down Simple Troubles: Four-Cycle

Finding the causes of troubles that keep an engine of yours from starting can be a fascinating (and money-saving) exercise. The first thing you should learn is to proceed by a logical sequence. Most people can't find out what's wrong easily because they don't follow any particular order in their troubleshooting. A haphazard approach wastes time and never leads you anywhere. This chapter contains a step-by-step method of testing that always works when followed exactly.

Reading this chapter, you'll develop your line of thought and sequence of tests that lead you right to any defects. You may find that repairing them still is a job for an expert, but you'll at least know what to tell him. The photos and words should also help you understand better the workings of your small engines.

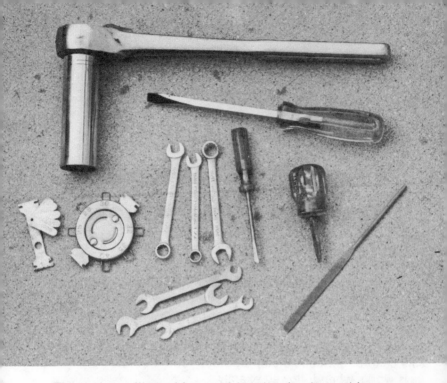

The tools you'll need for repairing the simple troubles are very few and inexpensive. You don't need the cabinets full of tools that professional mechanics use. Most hardware stores have tool kits. Some dealers carry little kit packets that contain all the necessary tools to care for your particular machine.

The tools you need for minor troubleshooting include a spark plug socket wrench, a feeler gauge, an ignition file, and a small screwdriver. The spark plug socket can be one of the stamped-steel kind; they're cheap and will do the job okay. The best kind of feeler gauge to buy has wires for "feeling" the gap on spark plugs and flat blades for the breaker-point gap.

Simple tools usually cost less at discount and hardware stores than at dealers' stores. But if you aren't sure about a particular tool, go to your dealer's. You'll pay the little bit extra, but he can equip you with exactly what you need.

Other tools that help you take things apart and get them back together: an adjustable end wrench (like Crescent), a pair of adjustable pliers of the kind sold by Channellock, a set of combination box and open-end wrenches, and screwdrivers of assorted sizes.

These tools should be of the highest quality you can afford, since they must withstand rough use. Even top-quality tools aren't expensive. You'll find them at dealers, auto-supply stores, or large retailers such as Sears, Roebuck or Montgomery Ward. The major tool manufacturers have a lifetime guarantee on their tools that justifies the extra cost.

One specialty tool is very helpful and relatively inexpensive. That's a pair of *snap-ring pliers*. The best kind available for the average handyman has interchangeable tips. You can buy them at most auto-supply or hardware stores.

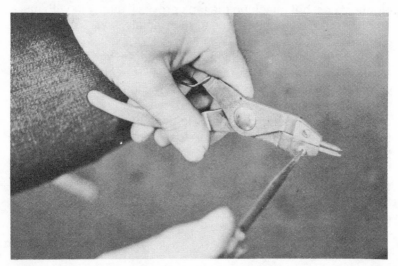

The tests shown on the remaining pages of this chapter apply to any small engine. What it's on doesn't matter—lawnmower, minibike, garden tractor, or whatever. You make the same tests whether the engine crankshaft is horizontal or vertical.

TEST 1. You may think this test silly and unimportant. But it's not. CHECK THE FUEL SUPPLY. Experienced mechanics cite repeated occasions when a customer swears the machine has plenty of gas, but checking the tank proves it empty or with just a tiny bit in the bottom. The customer lamely says, "Well, it had plenty the last time it was used."

Never take anyone else's word for anything when you're troubleshooting. The surest cure for "no fuel" is to fill the tank before you start the tests. Then there's no question.

If you don't have your fuel-storage can handy, go ahead and check the fuel level in the tank. Remove the cap and stick your finger into the tank. If your finger doesn't come out wet, then the fuel supply is meager. In a large tank, your finger might not be long enough to reach the liquid when the level is low but still adequate. Stick a wood ruler into the tank to measure the fuel level.

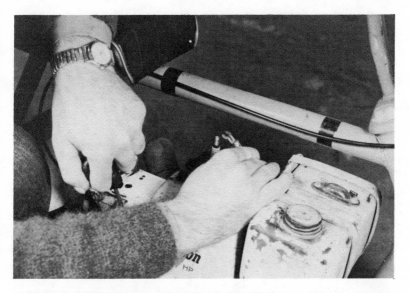

TEST 2. REMOVE THE SPARK-PLUG WIRE from the top of the plug. Hold the wire end about 1/8 inch from the top of the spark plug and crank the engine a couple of times. You should see a spark jump across the gap from the end of the plug wire to the top of the spark plug.

This checks operation of the magneto and the quality of the spark-plug wire. If no spark jumps, then check the wire itself for wear or breaks. Any decrepit-looking wire should be replaced as a matter of practical maintenance (even if spark is okay in this test).

Verify the connection between the plug wire and the coil. You will have to remove the flywheel cover and starter for this. Page 93 shows how to take this cover off. The wire must have a good solder connection where it attaches to the coil. Never just twist the connection and leave it. The electric current would eventually lose part of its power.

If the wire and the connection are okay, take the machine to a reputable mechanic to have the magneto and coil checked and repaired if necessary. Of course, if the spark jumps across the gap as described, proceed to the next test.

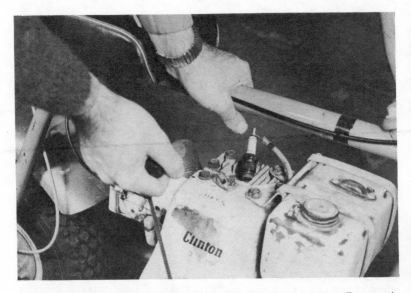

TEST 3. REMOVE THE SPARK PLUG from the engine. Reattach the wire to the top of the spark plug. Place the threaded end of the spark plug tightly against the engine head. Crank the engine again. Be careful that you touch only the insulation on the spark-plug wire. The electric shock from a small-engine magneto is sharp and jolting. You'll get it if you accidentally touch the bare terminal on the end of the wire as you're cranking the engine.

If the plug is good, you'll be able to see a spark jump across the electrode gap at the bottom end. A good hot spark is bright blue in color and makes a sharp, clear snapping noise. A poor plug (or weak magneto) generates a spark of yellowish color. Too, hardly any noise can be heard. This indicates that you need to replace the plug.

Refer to the owner's manual for the proper number for a new spark plug. (Someone else may have put a wrong one in by mistake.) Use only the spark plug recommended by the manufacturer.

Hold the new spark plug against the engine head. Crank the engine again. If the spark is good now, install the plug and proceed to next test. If a new plug doesn't make the spark snappy and bright blue, something is wrong in the magneto and needs the attention of a reputable mechanic.

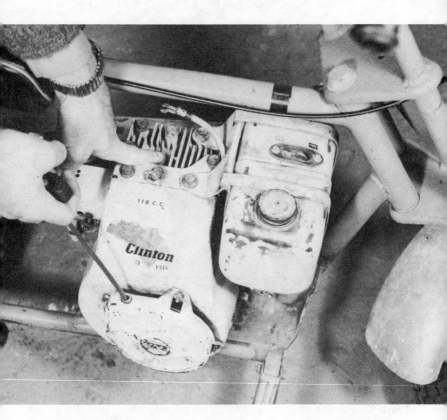

TEST 4. CHECK COMPRESSION. Leave the spark plug out of its hole. Place a finger or thumb over the spark-plug hole. Crank the engine so it turns over once. A piston that's developing good compression blows your finger out of the spark-plug hole. If it doesn't, something is amiss with either the piston rings or the valves. This condition must be remedied by your dealer or other reputable mechanic.

Be careful not to stick your finger down into the spark plug hole. It could reach into the combustion changer so far the piston might hit it at the top of the stroke. That could bruise or mash the end of your finger.

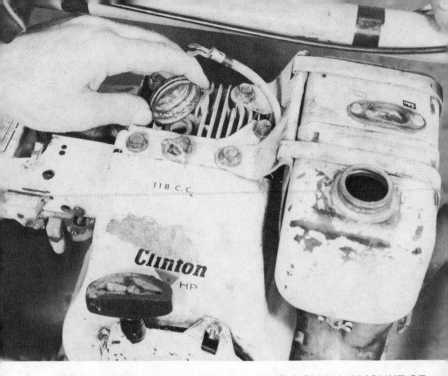

TEST 5. Engine still won't start? POUR A SMALL AMOUNT OF FUEL directly into the cylinder. Use the lid of the fuel tank. Pour a capful of gasoline through the spark-plug hole into the cylinder.

Immediately replace the spark plug and reattach the spark plug wire. Crank the engine rapidly a few times to start it. If the engine starts and then quits again, you know something is blocking the flow of fuel. The next tests help you find out what.

The engine may keep running. Yet, when you shut it off, you can't start it again without another capful of raw fuel. This suggests that some fault inside the carburetor keeps the proper air-fuel mixture from entering the cylinder. The additional vacuum, created by the engine's running instead of being turned slowly by the starter rope, is enough to keep the fuel mixture flowing once the engine is started. This indicates the carburetor needs a thorough cleaning inside—a job for your mechanic.

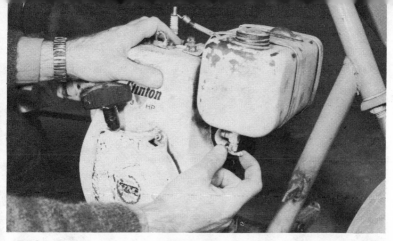

TEST 6. If the engine quits after a capful of gasoline in the cylinder has started it, the next step is to find where the fuel supply is blocked. DETACH THE FUEL LINE from the tank and see if fuel runs out of the tank outlet. If no gasoline comes out of the tank, remove the tank from the machine. Wash the tank out thoroughly with clean gasoline. This should remove all sediment. Don't stop flushing the tank out till you're sure it and the outlet screen are absolutely free of dirt and sludge. Replace the tank and refill it with clean gasoline. Fuel should flow freely from the tank outlet now.

TEST 7. Reconnect the fuel line to the gas tank. REMOVE THE OTHER END of the fuel line from the carburetor. Fuel should flow freely through the line. A trickle from the fuel line or no fuel flow at all means the fuel line is plugged. Remove it completely and blow through it to clear out any obstructions. Connect the line back to the tank and recheck for fuel flow. Put on a new line if the old one is even slightly deteriorated.

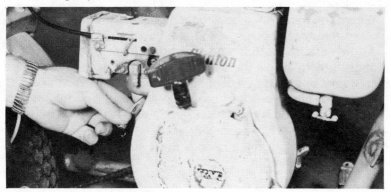

TEST 8. CHECK FUEL SUPPLY IN THE CARBURETOR. Some carburetors have a valve on the bottom of the fuel bowl to let you drain the carburetor (for winter storage). Push upward on this valve and gasoline should run out. If none does, something inside is blocking the flow of fuel in the carburetor.

Where the carburetor bowl doesn't have a drain valve, you'll have to remove the cover of the fuel bowl.

A few carburetors have no fuel bowl. Instead, the gas tank mounts directly below the carburetor. A small gasketlike rubber diaphragm, operated by a combination of vacuum from the top of the cylinder and pressure from the crankcase, siphons fuel up into the carburetor. A hose from the crankcase to the carburetor applies the pressure to accomplish this. If this type of carburetor doesn't deliver fuel mixture into the cylinder, someone must replace the diaphragm.

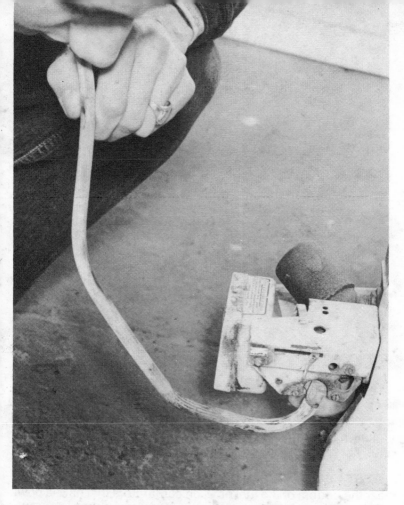

TEST 9. BLOW THROUGH THE CARBURETOR to clear out its internal passages. A small piece of sediment or dirt on a needle-valve seat can block the flow of fuel through the carburetor.

Attach a long piece of hose to the carburetor where the fuel line connects. A small piece of windshield-wiper vacuum hose suits this job just fine. Blow through the hose with your mouth. Never blow into a carburetor with compressed air. Too much pressure is likely to damage some of the internal parts or ruin a gasket.

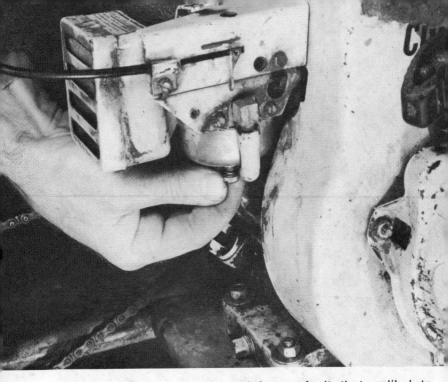

By now you have cleared most defects or faults that are likely to keep your small engine from starting. Now, using the sequence for cold engines described on pages 48 through 52, start the engine. If the engine is still hard to start, or doesn't run smoothly after you get it going, the air-fuel ratio probably is wrong.

TEST 10. THE FUEL MIXTURE ADJUSTMENT screw is located near the bottom of the fuel bowl on carburetors that have a bowl. Open the throttle about halfway, or to wherever it keeps the engine running. Turn the adjustment screw one way or the other *very slowly* until operation of the engine smooths out.

Carburetors with no fuel bowl have the adjustment screw on the side. The method of adjustment is the same, regardless of design. When you're adjusting the mixture ratio, never turn the screw very fast. Give the adjustment time to have an effect, and give yourself time to hear the change in the way the motor runs. A few new engines have these adjustments factory-set and sealed so you can't get at them.

TEST 11. The engine should be running smoothly at the Run setting of the throttle. Time to ADJUST THE IDLE SPEED. The screw for this almost always is located on top of the carburetor where the throttle cable attaches. With the throttle at its minimum or Idle position, you turn this screw gradually until the engine runs as slowly as it can without stumbling. If you set the screw to run the engine too slow, the engine may not accelerate smoothly when you open the throttle.

TEST 12. ADJUST THE IDLE MIXTURE screw. It controls the air-fuel ratio only when the engine runs at idle. Turn this screw clockwise *very gradually* until the engine starts idling rough, seeming to stumble or try to stop. Then turn the screw the other direction until the engine idles smoothly again. See if it will also accelerate smoothly when the throttle is opened quickly. You may need to reset the idle speed screw. You've found the best position when the engine idles smoothly and doesn't try to quit.

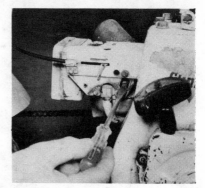

Some models of carburetor don't have separate adjustment screws for idle mixture and run mixture. On these carburetors, you adjust the mixture the same as on any other carburetor. However, you should verify that the engine runs smoothly at both idling and running speeds. Then adjust the idling speed as needed.

You should always go through the sequence of adjustments twice to be sure that all the adjustments are correct. If they are, you're finally ready to use the machine—maybe for a whole season—without trouble.

Chapter 5

Tracking Down Simple Troubles: Two-Cycle

Working with a two-cycle engine entails some thinking that is different. Because of design differences, a two-cycle engine runs at a higher speed than a four-cycle, to produce equivalent power. The addition of oil to the gasoline makes the condition of the spark entering the combustion chamber more critical.

The difference in carburetors affects starting a two-cycle engine. The carburetor, instead of mounting directly outside the cylinder head like on a four-cycle engine, mounts on the opposite side of the engine. The fuel mixture must therefore flow through the crankcase to reach the top of the piston—the combustion chamber—to be burned. The four-cycle fuel mixture flows directly into the combustion chamber.

But the tests by which you track down troubles change hardly at all. Two-cycle operation has the same basic requirements: a correct fuel mixture in the cylinder, and a good hot spark at the right time.

TEST 1. As before, CHECK FUEL LEVEL in the tank. You know all the ways. They are discussed on page 72. Better, fill the tank before you begin troubleshooting the engine.

TEST 2. A two-cycle engine in cold weather must be primed before you try to start it. PUSHING THE PRIMER BUTTON runs some raw fuel into the barrel of the carburetor where the air flows through, bypassing the jet. This lets the incoming air pick up more fuel on its way to the cylinder. The mixture entering the cylinder therefore is much richer. That makes starting a cold engine easier.

Keep pushing the primer button, before you crank the engine, until you see some fuel leaking out where the primer shaft enters the carburetor. If no fuel appears, continue the tests but expect to find carburetor or fuel-line blockage later.

TEST 3. SEE IF ANY SPARK reaches the spark plug. Make sure the kill switch is open. Pull the wire off the spark plug. Hold the bare wire end about 1/4 inch away from the top of the spark plug. Crank the engine. A spark should jump the gap from the wire end to the spark plug. No spark suggests you should take the machine to a dealer to have the magneto analyzed and repaired. A long, snappy blue spark tells you the magneto works properly.

TEST 4. TAKE THE SPARK PLUG OUT of the engine head and hook the wire back to it. Put the threaded base of the spark plug against the engine and crank the engine again. Spark should jump across the electrode gap at the bottom of the spark plug. If it doesn't, the spark plug is probably bad. Replace it with a new one of the type recommended for the engine.

The spark might be too weak to ignite the fuel mixture. A good spark is bright blue in color and makes that snapping sound. A poor spark is orange or yellow in color and lets you hear hardly any sound.

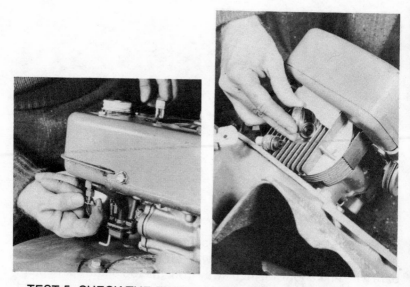

TEST 5. CHECK THE FUEL supply to the carburetor. It's common to forget to open the shutoff valve at the fuel tank.

TEST 6. While you still have the spark plug out, CHECK COMPRESSION. As described before, the easy way is to hold your finger tightly over the spark-plug hole and crank the engine once. Enough compression definitely blows your finger off the hole.

TEST 7. POUR A CAPFUL of gasoline directly into the cylinder through the spark-plug hole. Replace the spark plug immediately and crank the engine. That should start it. If it quits after a few seconds, the fuel supply is blocked somewhere.

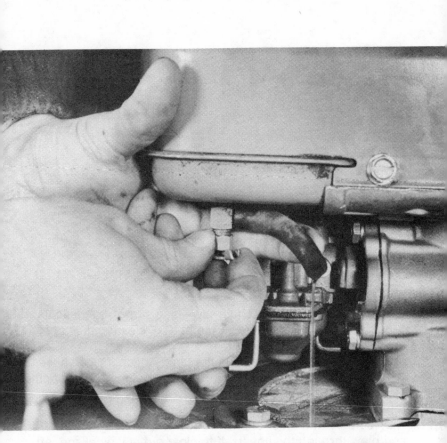

TEST 8. DISCONNECT THE FUEL LINE from the tank. A steady stream of gasoline should pour out of the tank when the shutoff valve is opened. If no gasoline or hardly any runs out, inspect inside the tank with a flashlight or electric light (no flames). Sediment in the bottom of the tank, even a tiny bit, can block the flow of gasoline through the strainer screen at the tank outlet. If you discern or even suspect any sediment, take the tank off the machine and flush it out thoroughly. Sometimes, merely shaking the tank dislodges the sediment enough to let the fuel flow again.

When you add gasoline, always be sure it is clean.

Taking the tank off of a chain saw—and some other machines —may be impractical. There's a method for clearing the outlet of any sediment.

Disconnect the gas line from the tank. Attach a clean hose to the tank outlet pipe. Blow back through the hose. That should clear any obstructions from the outlet. Gasoline should then flow freely from the tank in a steady stream.

Disconnect the other end of the fuel line from the carburetor, and blow through the line itself to clear out any blockage. After the line is reconnected to the tank, fuel should also flow through it in a steady stream.

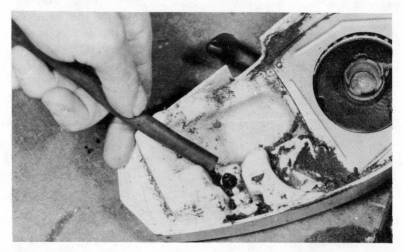

You can clear out the chain saw carburetor the same way. Attach a clean hose to the carburetor where the fuel line connects. Blow into the carburetor (by mouth, never with compressed air). The air rush clears out any sediment that has entered the carburetor and blocked the needle valves. A piece of sediment smaller than the diameter of a straight pin can close off a needle valve completely, since the opening is so tiny. A closed needle valve stops the process of mixing air and fuel, making the engine very hard or impossible to start.

Once the engine is running, adjust the mixture of air and fuel. Never mess with the mixture screws while the engine is stopped.

First adjust the run mixture to let the engine run smoothly at operating speed. Try loading it down (heavy grass for a mower) to make sure the engine delivers full power to do the work needed.

Next, adjust the idle speed as slow as the engine will run without stopping. Then adjust idle mixture so the engine idles the smoothest, yet will accelerate evenly and without hesitation.

The procedure for carburetor adjustment is the same for all two-cycle engines. However, the Jacobsen lawnmower engine has a wheel that adjusts the run and idle mixtures, instead of screws. The wheel turns a shaft that opens or closes off small holes down inside the carburetor. That controls how much fuel gets through the holes and into the carburetor barrel—where it mixes with the air. The one wheel sets the proper fuel-air ratio for both idle and run. That makes the setting critical; be sure idle and acceleration are both smooth when you finish.

Chapter 6

Minor Repairs You Can Make

When you buy a new piece of machinery powered by a small engine, treat it as you would a new automobile. Locate a dealer with a reputation for excellent service. Good dealers help anyone, whether you bought your machine there or not. That's the best place to buy any parts you need to keep your engine in top shape. Best of all, the service department can take care of repairs you can't do yourself or don't want to tackle.

Some owners like to take care of their own equipment and do all the minor repairs themselves. This chapter shows you such repairs as replacing points and condenser. You'll see how starter mechanisms come apart and how to replace a broken spring or pull rope. You can decide for yourself whether you want to get in that deep or not. Repairs more extensive than these, leave to an experienced small-engine mechanic. He has the special tools needed, and the special know-how.

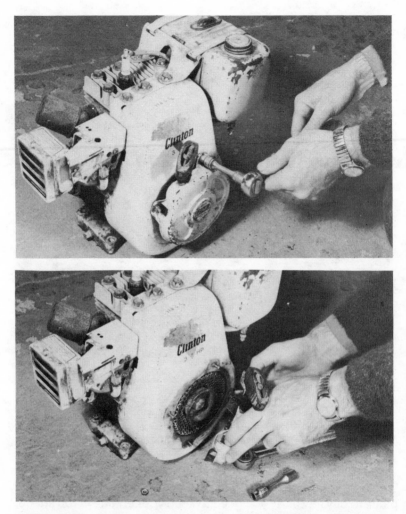

The repair your small engine is most likely to need, and one you can probably handle okay, is putting new points and condenser in the magneto. Just be sure you read and understand this entire chapter before you attempt this job.

Start off by removing the starter mechanism. Usually, it bolts to the shroud that covers the flywheel. The starter is spot-welded to the shroud of some engines. Look closely, as you remove the starter, to notice how the starter engages with the flywheel when the rope is pulled. Then you can replace it properly when you reassemble the engine.

With the starter off, the flywheel shroud comes next. That's the thin-metal cover that houses the flywheel and keeps leaves and trash away from it. Take out the bolts on the underside. Some of the bolts that fasten the shroud on top also hold the cylinder head on. You should reinstall them with a torque wrench—tighten to 20 ftlb—when you reassemble the engine. Watch carefully how the shroud comes off and clears the governor mechanism so you can replace it the same way.

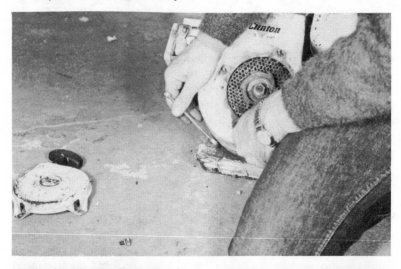

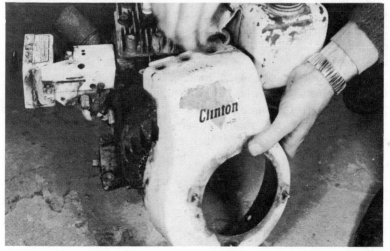

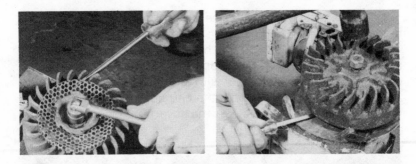

Disassemble the flywheel next. First remove the nut in the center of the flywheel. You'll have to hold the flywheel with a screwdriver in its fins in order to break the nut loose. However, handle the wrench with a steady, even pull. Do not jerk the wrench. The sudden yank might cause the screwdriver to break off some cooling fins and ruin the flywheel. The perfect balance built into it would be destroyed if any part were broken off. The unbalance would cause excessive vibration when the engine is running.

After you remove the nut and sawtooth-looking catch for the starter, replace the nut. Screw it down until the outer side of the nut is flush with the end of the shaft. Now put the screwdriver under the flywheel and apply slight upward pressure. Holding the screwdriver like that, tap the nut sharply with a hammer. This ought to loosen the flywheel. The flywheel mounts on a tapered end of the crankshaft. The taper fits so tightly you need the hammer-tap to jar them apart. The nut flush with the end of the crankshaft keeps the threads undamaged when you hit the end with a hammer.

After you pop the flywheel loose from the crankshaft, remove the nut and the flywheel. Watch out and don't lose the key that keeps the flywheel locked so it can't spin except *with* the crankshaft. It's made of a special nonmagnetic material; if it gets lost or damaged, buy an exact replacement from the dealer.

The flywheel, as is described earlier on pages 21 and 39, carries one half of the magneto. Permanent magnets mounted in the side of the flywheel create the magnetic field necessary to generate an electric pulse. A coil mounted on the engine block makes up the other half of the magneto. Each time the magnets in the spinning flywheel pass by the coil, their magnetic field cuts through the copper-wire windings of the coil. The moving magnetic field induces a surge (pulse) of electricity in the coil.

The coil has two windings. One of them, the primary winding, develops that initial electric current directly from the magnets. This winding, together with the breaker points and condenser, forms the ignition system's *primary circuit.*

The second winding, of very fine wire, has many more turns of wire than the primary winding. This winding and the spark-plug wire form the ignition *secondary circuit.* The coil acts as a step-up transformer, increasing the voltage pulse from the primary circuit to a voltage high enough to jump across an air gap (at the bottom of the spark plug). The spark-plug wire, attached to the end of the secondary winding, carries the high-voltage electric pulse to the spark plug.

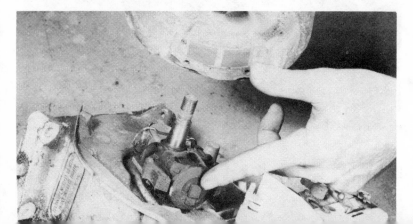

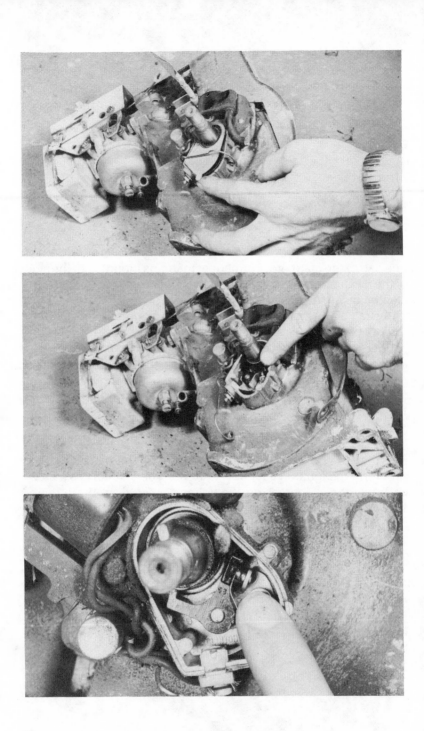

The breaker points provide a carefully timed switch. One contact connects to the primary winding of the coil; the other is grounded to the engine block. When the points are closed (contacts together) all of the electric pulse generated by the magneto flows through the points to ground. This actually prevents the pulse from being coupled to the high-voltage secondary winding.

At a specific point in the flywheel's revolution, an eccentric lobe on the crankshaft pushes the points open. This has been designed to occur just as the piston passes top-dead-center at the end of its compression stroke. When the points open, the pulse suddenly is allowed to develop in both primary and secondary windings. A great surge of electric current, at very high voltage, passes from the secondary winding to the spark-plug wire and the spark plug. The current jumps the electrode gap of the plug, and the ensuing spark ignites the fuel compressed in the combustion chamber.

The coil, by its transformer nature, generates a heavy counter-surge when the spark jumps. This "reverse" voltage goes back to the primary circuit and still-open points. This counter-surge of electricity must be absorbed by the *condenser*. Without it, the point contacts would soon be melted away by arcing. A defective condenser causes the point contacts to become burned and pitted in only an hour or two of operation.

The voltage step-up in the secondary winding occurs because the primary winding has many fewer turns of wire than the secondary winding. The voltage increases to approximately 20,000 volts. It must, so it has enough push to jump across the electrode gap at the bottom of the spark plug.

By the time the high-voltage pulse has jumped across the gap, the breaker points close again. The primary circuit is thus abruptly grounded again. All voltage, current, and spark ceases until next time the magneto develops a surge and the points open and allow it to be applied.

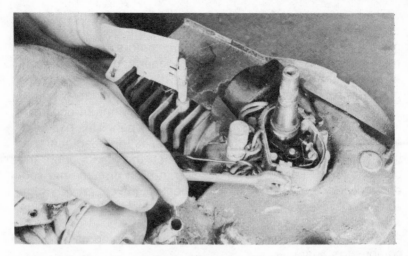

You need to replace the points and condenser at least once a year or every 50 hours of operation. Worn points and condenser constitute one major cause of hard engine starting.

Once you remove the flywheel, you'll see the dust cover over the points. A spring clip across the top holds the cover in place (page 96). Be careful not to damage the gasket underneath this cover. The gasket keeps dirt away from the points. Contaminants cause them to burn prematurely.

Now take loose the wires that run from the coil and condenser to the points. Remove the screws that hold the points in place.

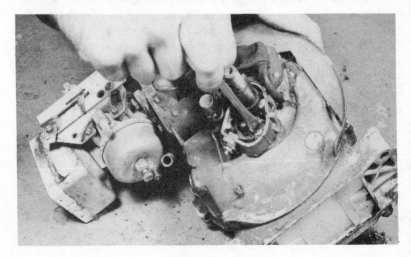

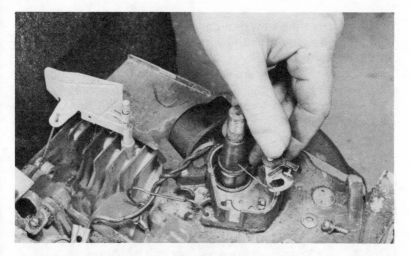

Lift out the points. Check the new set to be sure they are identical. There have been cases of mislabeling or a mistake in reading a part number. You don't want the wrong type of points in the machine. This quick check prevents such grief.

Install the new points. Leave the mounting screws slightly loose until you set the points for the proper breaker gap (a bit later).

Next remove the screw or nut that holds the condenser body in place. (You already took the wire loose, along with the others.) Lift the condenser out. Also check the condenser to make sure the old one and new one are alike. Then install the new condenser and tighten the mounting screw. Reconnect the condenser wire and coil wire to the terminal on the points and tighten both securely.

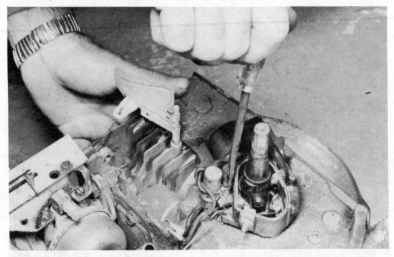

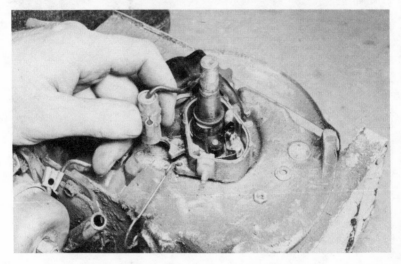

Now you're ready to set the gap that exists between the contacts of the points when they're open. The breaker gap setting for most small engines is .020 (20 thousandths of an inch). The gap setting is critical to efficient engine operation and to easy starting.

The points open when their rubbing block rides up on the highest point of an eccentric lobe on the crankshaft. To find the highest point on the lobe, remove the spark plug (so there's no compression) and turn the crankshaft (by hand) with the mower blade or the other end.

Slip the blade of the feeler gauge into the gap. The blade should touch both contact faces, but with the slightest touch possible. Move the free breaker arm so that contacts barely touch the feeler gauge. Then tighten the mounting screws. Recheck the gap to insure that nothing moved while you tightened the screws.

Rotate the crankshaft a few times and watch to see that the points open and close properly. Then find the high point on the lobe and check the point gap one last time. Nothing should have changed while the crankshaft was being turned, if you tightened everything down snugly.

Before you cover the points, look over all the connections. They must be tight. Anyplace there's a loose connection, the circuits won't be made properly; points and condenser can't function. Also, the surges of electricity might arc in a connection and burn a wire loose completely.

Replace the points cover and its gasket. Seat the gasket carefully so you don't damage it. After you replace the cover, turn the engine crankshaft a few times by hand to be sure that nothing is rubbing.

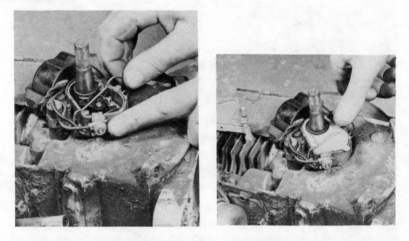

Replace the flywheel, and tighten the nut that holds it on. Pull it to 40–45 ftlb with a torque wrench. Put the shroud and starter back on. Tighten those bolts in the shroud that also hold the head on to 15–20 ftlb of torque. Be very careful to tighten any other loose head bolts to the specified torque.

The magneto on a two-cycle engine is very similar to the one on a four-cycle engine. A principal difference lies in the cooling fins on the flywheel. Nevertheless, operation is the same. The points in a two-cycle engine work twice as hard, since the plug ignites the fuel every time the piston passes top-dead-center instead of every other time as in a four-cycle engine. You use the same procedures to replace the points and condenser on a two-cycle engine.

If you decide the coil is bad and must be replaced, use the procedures already described. After you replace the coil and put the flywheel back together, you have to set the air gap between coil and magnets. The mounting bolts let you shift the coil into the right position. The proper gap is .020 to .025, and must be held. Some coils have two pickup faces instead of one. Take care to set the gap the same for both faces. If the air space is too wide or too narrow, the magneto can't produce enough primary voltage to make a spark (secondary voltage) that will operate the engine.

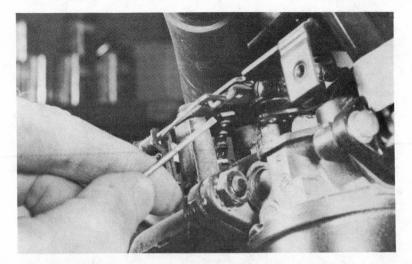

One trouble that's not uncommon involves the kill mechanism. Sometimes it malfunctions and the engine can't be started or can't be stopped.

Most small engines use a wire-and-leaf type. The wire connects to the ignition coil primary on one end and to an insulated terminal on the other. The terminal is mounted near the carburetor. A thin metal leaf connects mechanically with the throttle linkage on the carburetor. When you pull the throttle all the way back, it moves the leaf to touch the terminal and wire. The leaf thus grounds out the electric pulse the coil would otherwise generate. No current reaches the spark plug, and the engine dies.

To ferret out the trouble in a bad kill mechanism, first disconnect the wire from the insulated terminal. Start the engine. Then touch the bare end of the wire to any part of the engine block. The engine should die. If it doesn't, inspect the wire; it's probably broken somewhere. If the wire kills the engine ignition normally, suspect the leaf or its links to the throttle cable. Make sure the leaf touches the terminal when the throttle is pulled all the way back. If it doesn't, readjust the throttle cable until the leaf does touch the wire terminal. The leaf may twist slightly; tighten it up where it's mounted or replace it with a new unit.

A case of "no ignition spark" can sometimes be traced to a shorted kill system. Inspect the insulation on the wire thoroughly. It may be cracked or burned someplace. If the wire is bare and touching metal anywhere, the coil is rendered inoperative. Replace the wire. Check the insulated terminal. It should have no cracks or looseness. Neither it nor the wire must touch anything metal when the leaf is away from the terminal. If the terminal is broken or an insulator cracked, replace terminal and insulator.

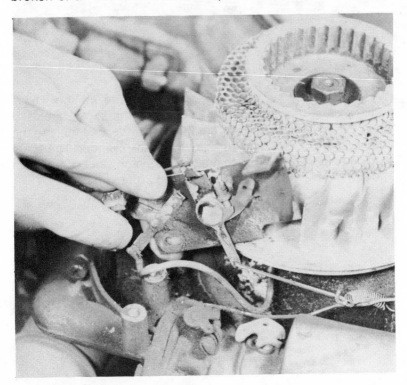

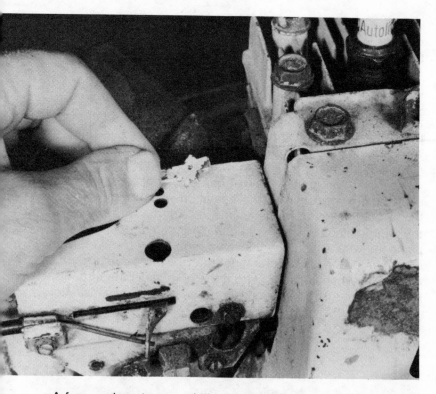

A few engines have no kill mechanism for the ignition. These engines must then be choked to death. If the choke is closed when the engine is running, that blocks off the air supply to the engine. A too-rich fuel mixture goes into the cylinder. The spark plug can't ignite a mixture that rich. The piston can't compress it enough to let it ignite. The engine stops running.

Chapter 7

Repairs to Starter Systems

You'll see more *recoil-type* starters than any other on small engines. This starter has a pull rope coiled around a pulley. Both are enclosed by a housing, although the rope end protrudes through a guide slot or hole and terminates in a handle. A spring hooks the pulley. When you pull the rope out of the housing to crank the engine, the spring winds up tight. When you release the rope, the spring force turns the pulley, recoiling the rope around the pulley.

You'll find a few variations in how the starter mounts to the flywheel shroud. But the recoil mechanism is basically the same in all of them. The manner of engaging the starter pulley to the flywheel when the rope is pulled varies too, yet all versions work very much alike. Replacing a broken rope or a defective rewind spring are the only repairs you should undertake on this type starter.

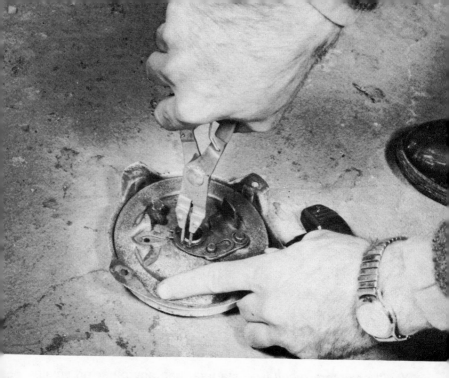

Repairs necessitate removing the starter mechanism from the engine. Page 92 shows how to accomplish that. After you remove the starter from the flywheel shroud, you'll see a snap ring directly in the center of the starter unit. It holds the pulley in place on the center shaft of the starter housing. You'll need a pair of snap-ring pliers (page 71) to remove this ring during repairs.

With the snap ring removed, you can disassemble the starter. First remove the half-moon or crescent-shaped slide that mounts on the center shaft. The function of this slide is to move the pawl (beneath the slide) out to engage the flywheel when you pull the rope.

Now you have access to the pawl that fits under the slide. A C-clip holds it to the pulley face. A pair of sharp-nose pliers can help you remove the C-clip, or you can snap it free of its post with the tip of a small screwdriver. Lift the pawl free of the post.

Pay close attention to the sequence in which you remove the separate pieces. You'll reverse the order of removal when you reassemble the starter.

If the rope is broken, pull out the remaining piece through the hole in the face of the pulley. A new rope must be close to the same length and diameter as the old one. Some starters use a rope of larger diameter than others. The starter spring can't recoil the rope properly if the new rope is much larger than the old one. A smaller rope won't lay right on the pulley.

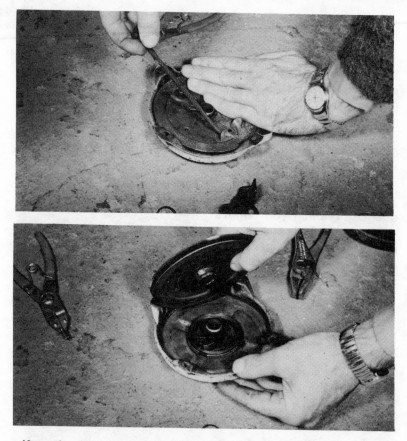

If you have to replace the spring, lift out the pulley. The spring has a loop on the end to hook the pulley. As you start to lift the pulley clear, turn your face away. Spring tension (there should be none) caused by some defect could snap the spring out into your face. This is unlikely, but play it safe.

Remove the spring and replace with the new one. Take care that the new spring lays so its direction of winding is the same as the old one.

Hold the spring in place in the housing with one hand. With the other, turn the spring until the outer loop has engaged the stop molded into the housing.

Now replace the pulley on the center shaft. Watch that the inner loop of the spring engages the tab on the pulley. Once you have the pulley back in position, hold it there as you replace the pawl and the half-moon slide. The pawl, as you recall from when you removed these parts, goes next to the pulley. The one C-clip holds both parts in place.

Replace the snap ring that holds the pulley to the housing. Jiggle the pawl and slide a few times to assure yourself that they move freely. If necessary, you can smear a very small amount of silicone grease on the slide to aid smooth operation.

You're ready to install the new rope as soon as you have the starter reassembled. Rotate the pulley until it can't be turned any further, to wind up the spring. From here on, the pulley remains under spring tension. You'll have to hold the tension with one hand and execute the rope-threading with the other.

Back off the pulley one turn, to where the guide hole in the housing lines up with the rope hole in the pulley. From the pulley side, thread the rope through both holes. Remember to hold the pulley with the spring wound tight. You'll find it's much easier to start the rope from the pulley side. As you push, twist the rope at the same time. A twisted rope becomes solid and is much easier to push. If the going gets awkward, you can use a small screwdriver to help cram the rope through the holes.

When you finally have the rope threaded through both holes, leave about equal lengths of the rope hanging out of each. Release the pulley slowly, letting it coil the rope around itself.

Leaving the ends hanging as you release the pulley will jam the rope and hold the pulley steady while you knot the ends.

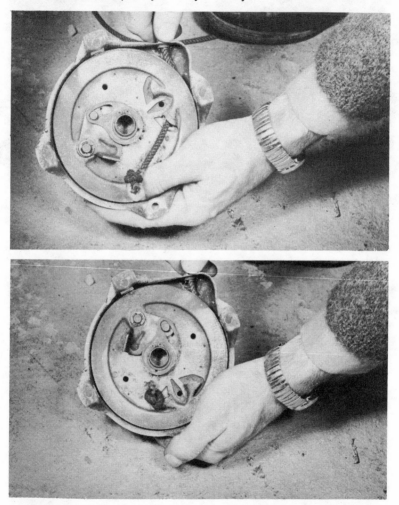

Tie a knot first in the end of the rope that sticks out of the hole in the pulley. Pull the knot as tight as you can get it. If the rope is nylon, take a match and melt the end of the rope into the knot. This trick prevents the knot from coming untied as you jerk on it later whenever you start the engine.

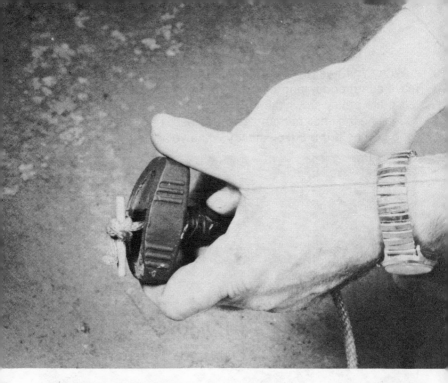

Thread the pull handle onto the end of the rope that sticks out of the starter housing. Tie a knot in the end of the rope. The knot must wrap around the metal pin that fits inside the handle. Pull the knot as tight as you can get it. Melt the end of a nylon rope into its knot. Slide the handle up to the end of the rope, and slip the pin into place in the handle.

Now backwind the pulley and work the entire rope through both holes until the restraint knot rests tight against the pulley. Finally, pull the handle and rope all the way out. Give a few sharp tugs to seat both knots and to assure that neither can come untied when you use the starter.

Install the starter back on the flywheel shroud. As you slide the starter into place, ascertain that the pawl meshes with the serrated drum on the flywheel. The pawl catches on one of those serrations when you pull the starter rope. The circular motion you impart to the pulley as you pull the rope turns the flywheel and thus the crankshaft. Once the starter is bolted securely in place, pull the rope all the way out a few times. That'll tell you whether the pawl engages the starter drum on the flywheel as it is supposed to.

Bolt up the rest of the shroud and you're finished. Repairing a recoil starter assembly isn't such a huge job.

The recoil starter on a small chain saw operates much like those on other engines. The housing for the starter assembly also contains the gas tank. Remove the housing and tank by first removing all the screws that hold the housing to the engine. Note the lengths of the screws. Where they differ in size, you'll have to replace them in the proper holes.

Pull the fuel line loose from the carburetor and plug the free end with a bolt, so no fuel spills while you work on the starter.

Remove the inside cover. That exposes the pulley and spring, in case either has to be replaced. The procedure starting on page 110 works fine if you have to replace the recoil spring.

When you've completed the repair, replace the inside cover.

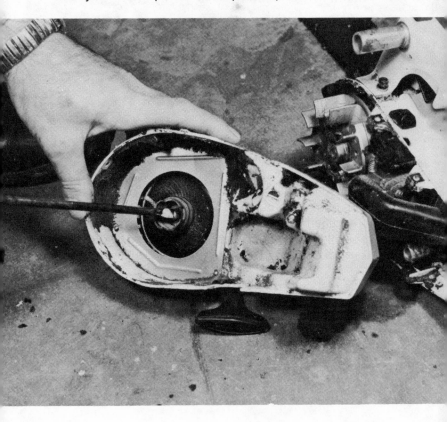

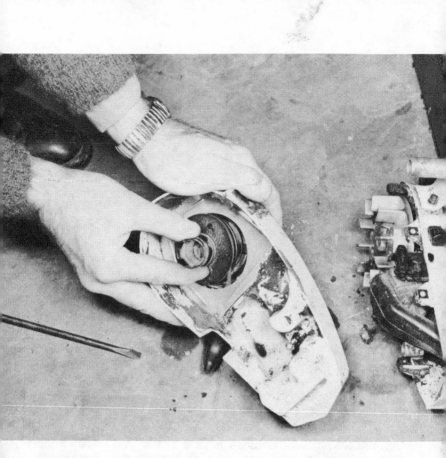

To replace a broken starter rope, remove the inside cover so you can get at the pulley. Try to extract the remaining piece of rope without completely removing the pulley. You don't want the spring to come unwound.

Thread the new rope into the hole in the pulley and coil all of its length around the pulley. Wind up the pulley about three turns, so the rope will have some tension on it when the starter is reassembled and the rope is not being pulled.

Push the free end of the rope through the hole in the starter housing. Pull out enough rope to hold easily while you install the handle and tie a knot in the end. Then push the pulley firmly into place and hold it there until you get the handle on. Tie the handle knot. Replace the inside cover and the assembly is ready to be put back on the saw.

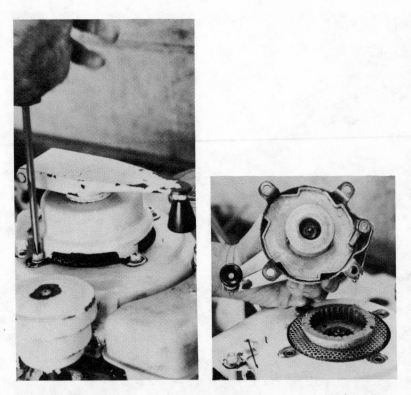

Some lawnmower engines have a crank-type *impulse* starter. The starter fits onto the flywheel shroud. The spring works directly against a flywheel drum to spin the engine for starting. You use a fold-down crank on top of the starter to wind up the spring. A spring-loaded catch, mounted on the side of the shroud, holds the flywheel stationary while you wind the spring tight. After the spring is tight, you fold the crank down and pull the catch free. That releases the spring tension and spins the flywheel. On some, just folding the crank releases the tightened spring automatically.

Anytime you have a problem with this type of starter, take the machine to a dealer for repairs. It takes special tools, parts, and knowledge to repair this type starter. The tools cost more than you'd pay a dealer to repair the unit.

Chapter 8

Propulsion Mechanisms

The drive train for one self-propelled lawnmower consists of a V-belt drive and a chain drive. A V-belt from a pulley on the end of the engine crankshaft drives a pulley on an idler shaft. A sprocket chain, similar to but heavier than a bicycle chain, carries drive power from the idler shaft to the rear-wheel axle.

Variations of this system abound. Positioning of the drive train in relation to the engine and the axle may differ. Some small models don't have V-belt or chain, but a simple system of gears. V-belt and chain withstand heavy use better.

Riding mowers usually differ slightly because they have a transmission. It may be a small one-speed forward and reverse transmission, or it may have three or more speeds forward and one reverse. In either case, the drive is similar. Typical is a V-belt drive from engine to transmission and chain drive from transmission to axle. The belt normally rests loose on the pulleys. When you push the clutch pedal or lever, a rod pulls an idler pulley against the belt and tightens it, making the rear wheels turn. When you release the pedal or lever, the belt goes slack and the mower stops.

Here is more detail. This mower has a shaft that the engine crankshaft turns all the time. A pair of bevel gears apply crankshaft power to the shaft. The end of this shaft that you see has a V-belt pulley secured on it. The pulley turns as the shaft turns.

A second pulley mounts on an idler shaft behind and slightly above the drive shaft. In the machine pictured here, the idler shaft is movable. You merely move the shaft and pulley farther from the drive pulley to tighten the belt and drive the mower.

The idler shaft also has a small sprocket wheel mounted on it. A larger sprocket wheel on the rear axle is directly in line with the one on the idler shaft. The two sprocket wheels are connected by a roller chain slightly larger than a bicycle chain. The two sprockets are different in diameter so the rear axle turns slower than the drive shaft from the engine. If it didn't, you couldn't keep up with the mower.

Each sprocket is held in place by a split pin that extends through the shaft and the sprocket hub. If a sprocket wears down or gets a tooth chipped and needs replacing, you drive the split pin out of the shaft and sprocket hub. Then you can slide the sprocket off the shaft and install a new one.

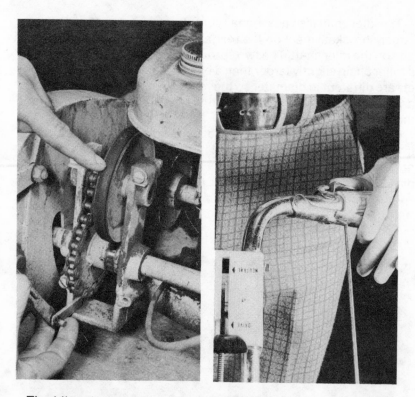

The idler-shaft bracket is pivoted. You move it to tighten and release the V-belt. A control rod from the bracket to a lever on the handlebar of the mower lets you engage or disengage the drive mechanism at will.

The rod end next to the bracket is adjustable. You set it so the travel of the lever pulls the bracket just far enough to disengage the V-belt from the pulleys.

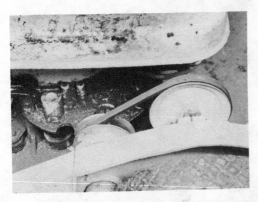

When you pull the lever against the handle of the mower, that motion pulls the bottom of the idler-shaft bracket backward. The top of the pivoted bracket, where the idler pulley is mounted moves forward. The belt goes slack. The drive train disengages.

A thumblatch on the mower handlebar slides a pin through a hole in the drive hand lever. The pin holds the lever in its Disengaged position.

When you want to engage the drive, you pull the lever against the handle and push down on the thumblatch on top of the handlebar. The pin comes out of the hole and the hand lever is released. The idler bracket moves in a direction that tightens up the V-belt. A spring holds the bracket in this position to maintain tension on the belt.

The chief component in the drive train of a minibike and a chain saw is a *centrifugal clutch.* When you speed up the engine, the clutch tightens and turns the rear wheel (minibike) or the blade (chain saw).

A centrifugal clutch usually mounts directly on the end of the crankshaft. It's held in place by two things: A square key fits a slot in the crankshaft and makes the center portion of the clutch turn with the crankshaft. The other, a set screw, keeps the clutch assembly from sliding off the end of the shaft.

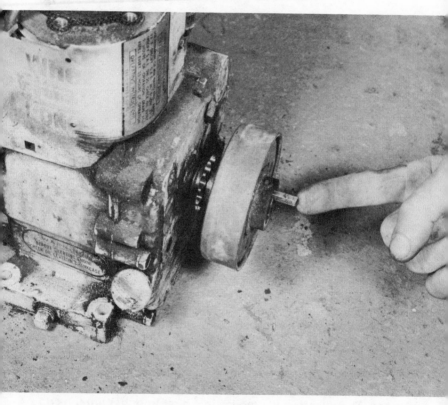

This kind of clutch, as you might guess, works on a principle called *centrifugal force*. When you speed the engine up, the crankshaft spins the center hub of the clutch mechanism. Two shoes are mounted opposite each other on the center shaft. The two shoes are forced outward against an outer drum by the forces developed by spinning the hub. This outward pressure applied by the spinning is centrifugal force.

The spinning shoes exert tremendous pressure against the outer drum. It begins turning and quickly develops the same speed as the inner hub and shoes. They turn as one. The drum transfers power to the wheels through a chain or V-belt to pull the vehicle, or directly to the blade to do the work required.

Most centrifugal clutches wear out the drum and shoes. The clutch begins to slip at all engine speeds instead of pulling the load placed on it.

Some clutches you should replace completely when this happens. They are fairly inexpensive. They have no lining material on the shoes. The inside of the drum and the outward face of the shoes become scored or excessively slick from the spinning of one metal face against another.

A more expensive version has a lining on the shoe face. It's of the same material as the linings on automobile brakes and clutches. With this type of clutch, you can replace the shoes and make the clutch like new again.

The inside hub and the outer drum of the clutch are held together by a snap ring. You can take the two halves apart for repair simply by removing the snap ring.

The two shoes are held in place around the inner hub by a pair of springs. To replace the shoes, unhook the springs and set the shoes off the hub. Then hook one spring to both new shoes and position them over the hub properly. Then hook the other spring to hold them in place. Reinstall the hub-and-shoe assembly in the drum. Replace the snap ring.

The only other part of the clutch that can wear out and need replacing is the bronze bushing. It is pressed into the center of the outer drum. It forms the bearing surface between the two spinning halves of the clutch: drum and hub/shoes.

When the bushing becomes worn, the fit between the two halves of the clutch grows loose. You can hear the clutch rattle during operation. The clutch also slips some and lets the engine race when a load is applied. You'd best take the clutch to a dealer to have a new bushing pressed into place with special equipment.

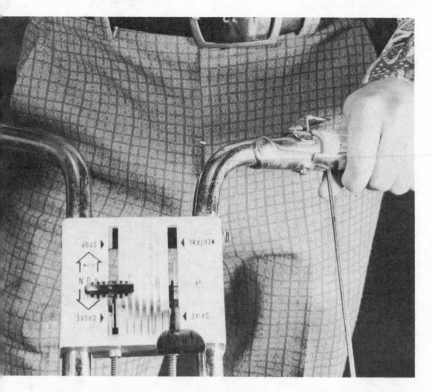

A few self-propelled walking mowers have a separate control to disengage the blade while you start the engine. The blade generally is driven by a V-belt that runs from a pulley on the crankshaft of the engine to a pulley on the shaft the blade mounts on. A third pulley, a movable idler, is controlled by a cable that runs to the handlebar of the mower. When you are ready for the blade to turn, you pull the cable handle to the "Cut" position. The idler pulley comes in against the V-belt. The belt tightens and the blade starts turning. This belt-drive mechanism is independent of the drive for propulsion.

Almost all riding lawn mowers include this feature. It would be almost impossible to start larger engines by pull rope if the blade were in gear all the time. This also qualifies as a safety feature. You can start the engine, get well clear of the blade, and then engage it safely.

Chapter 9

Save Dollars by Preventive Maintenance

Thoughtful care of a small engine prolongs its life and helps make most of its use troublefree. If you service your small engines as regularly and thoroughly as your automobile engine, your machinery could last many years.

After a summer's work, your small engines need winterizing for storage. The engine survives the nonuse better if you take certain steps. Too, it will start easier in the spring if you prepare it properly for winter idleness.

Take care of any minor repairs that were bypassed while you kept the machine busy. Most dealers are not as busy during the winter as in the spring; they can often get at repairs sooner. A few dealers even reduce repair prices for the winter months. Bring in your equipment then, instead of waiting until spring when everyone wants service posthaste.

Clean your lawnmower from top to bottom. Scrape all mud and grass from the underside of the mower housing. A screwdriver works, but a putty knife or other flat object is better. Without this scraping down, the grass and mud very soon rusts out the housing.

Wash the whole machine down with a stiff-pressure water hose. This cleans away dirt and debris that catches in the small cracks and holes in the machine. After this kind of thorough hosing, you may have to wait a bit until the magneto dries out before you can start the engine. Be sure to run the engine a few minutes after the washing, to warm it up so it dries thoroughly inside and out.

An excellent rust preventive trick: Clean the underside of the mower absolutely, with a wire brush, and coat the whole area with rust-inhibitor paint. Apply paint to the steel housing every winter, and it will last almost forever.

Take the blade off and have your dealer sharpen it. The blade needs this expert attention at least once a year. Your dealer can balance it then. During the summer, you can touch up nicks in the blade with a file. But balance should be checked professionally. Unbalance brings on excessive vibration, which wears out crankshaft bearings surprisingly fast.

After the blade has been sharpened, bolt it back on the mower securely and coat it with a light film of heavy oil or grease. This prevents rust and keeps the blade sharp longer when the mower goes back into service next year. As a precaution, before you start the mower for the first time every spring, retighten the bolt or bolts that secure the blade.

This is the time to clean the fuel tank thoroughly. Pour out all the unused gasoline and flush the tank a couple of times with clean gasoline. Shake it well. Flushing cleans out all the sediment and trash that accumulates in the tank during the summer. Leave the cap off awhile so what gasoline can't be poured out can evaporate.

Then start the engine and run it until it quits. This empties all of the gasoline from the carburetor and fuel line and prevents the carburetor gumming up while it sits through the winter. Gasoline that is allowed to evaporate from inside the carburetor leaves a gummy residue because certain additives in gasoline won't evaporate.

Remove the spark plug from the engine head. Pour about 2 oz of oil into the cylinder. Using the pull rope or your hand, turn the engine slowly a few times. This coats the cylinder wall and piston with oil. It prevents the piston rings from rusting to the cylinder wall while the engine isn't being used.

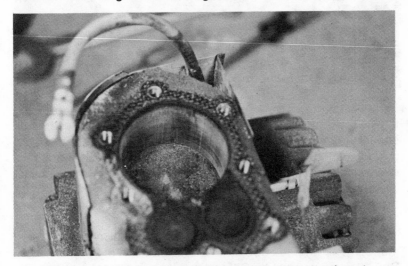

If you omit this important storage step, you stand a strong chance that the cylinder wall will be scored by the rings when you start the engine next spring. That happens because the rings and cylinder wall become completely dry from sitting idle and unlubricated over the winter months. Add the oil before you store.

Finally spring arrives. You should take time for a bit of preventive care as the mower comes out of storage. For example, change the oil before you start the engine. During the storage period, moisture accumulates in the crankcase; variations in temperature during the winter cause some condensation inside the engine. You should change the oil after every 25 hours of running anyway. And always check the oil level before you start the engine, EVERY TIME you use the machine.

The type of oil for your engine is specified in the owner's manual. The engine should have SAE 30 oil in the summer and SAE 10 if you use it in the wintertime. If you encounter extremely hot weather, change to SAE 40 oil for that period. The numbers refer to "weight" (actually density) or viscosity of the oil. Never use oil heavier than is necessary for the weather temperature, as heavier oil doesn't lubricate well.

Invest in a new spark plug in the spring. Besides being worn from last season, the old plug probably accumulated some rust through the winter months. Rusty electrodes on the spark plug can't develop as hot a spark as you'll get from a new one. Be sure always to install the type of spark plug the manufacturer recommends for your engine.

Check the points, and replace them unless the contact faces are bright and clean. New points and condenser every spring prevent a lot of trouble during the summer months. As a bonus, the engine uses less fuel per hour if the points, condenser, and spark plug are replaced every season. With the petroleum shortage and the rising price of gasoline, that saving alone can pay for the parts. And the relief of knowing the engine has a fresh beginning for the summer makes using the machine a happier experience.

Always pour fresh gasoline into the fuel tank in the spring. Throw out any that was stored all winter; it will have gums separated that are bad for the carburetor. Be sure the container you buy and keep gasoline in is absolutely clean inside.

It's dangerous to store gasoline over the winter. Besides deteriorating through evaporation, leaving a remainder that will not burn well, gasoline gives off fumes that can build up in a room and lead to explosion or fire. (Remember, it's gasoline vapor that explodes, not the liquid.) Never keep the gasoline in the house or basement, even in summer. Always keep the gasoline can, plainly labeled, in a well ventilated place.

If your engine is a two-cycle, always mix new gasoline and oil in the spring. Don't mix more than a couple of gallons at a time. You can't successfully store mixed gasoline for any long period of time. The gasoline evaporates and leaves a mixture that has too much oil. That makes the engine harder to start, it will smoke excessively, and the spark plug will foul in a very few hours of operation.

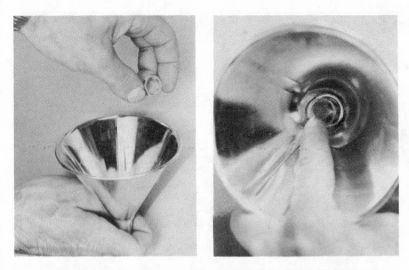

Any gasoline can carry a few impurities in it, even right from the service-station pump. Always use a funnel with a strainer screen to fill your fuel tank. If you can't find such a funnel to buy, make one yourself. Solder (or have it done) a faucet strainer screen in the funnel. The screen can be purchased at any hardware or plumbing store. It and a small funnel are very inexpensive. If you don't own a soldering gun or iron, a neighbor that has one could solder the screen in the funnel.

The air cleaner makes a lot of difference in how easily your engine starts, and how economically it operates. Air finds it difficult to get through a dirty filter. The carburetor can't provide the right mixture of fuel and air to let the engine run efficiently and properly.

To clean the air filter, which you should do each spring, first remove the screw that holds the air cleaner to the carburetor. Take the entire unit off the engine.

Take the bottom half of the housing off and remove the spacer tube that's in the center of the filter. Then take the foam rubber filter out of the housing pan so you can clean it.

The filter should be cleaned this way after every 10 hours of operation as well as every spring before you start the engine. Clean the filter more often during the summer if you use the engine where it's very dusty and the filter becomes clogged faster.

Probably the simplest way to clean the filter is to wash it well in fresh, clean gasoline. After the washing, you have to dry the gasoline out of the filter thoroughly.

Then soak the filter completely with clean oil. The foam rubber picks up any dust particles coming into the carburetor intake with the air. Dust particles trapped in the filter are kept there by the oil; they can't enter the engine and wear the carburetor or cylinder.

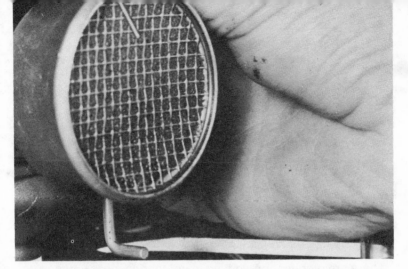

Naturally, not all air cleaners are built exactly alike. Some use filter material other than foam rubber. One material is synthetic horsehair. The hair-like material is woven tightly and held together by a screenwire covering. All this fits inside the air cleaner housing. Horsehair filters also must be washed with gasoline and soaked with oil, the same as foam rubber filters. Use an identical timetable (every 10 hours and every spring) for servicing hair filters.

Foam rubber is used almost exclusively in the air cleaners on chain saws. Cleaning the air filter on a chain saw is even more important than on a lawnmower. The saw itself creates an enormous amount of dust—sawdust.

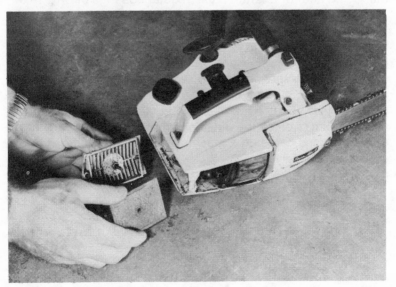

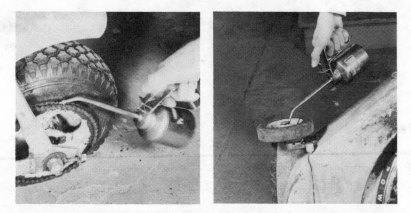

There are plenty of preventive maintenance ideas you can use to lengthen the life of your small engine. Your owner's manual describes several. Proper lubrication of moving parts outside the engine increases the life of any machine considerably. The drive chain on a minibike or lawnmower should be coated with oil often enough to keep the chain slightly damp all the time. Use only SAE 10 engine oil on chains. Heavier oil or grease won't penetrate inside the rollers properly, the pins that connect the links of the chain then wear out prematurely. Too light an oil merely runs out and/or evaporates in warm weather. Before you store the machine for the winter, coat the chain all over with light grease or heavy oil. This prevents rusting. But wash it off with gasoline in the spring and for operation use regular coatings of SAE 10 oil.

Oil the wheels on any rolling machine once a month during the operating season. If you prefer grease instead of oil, because it lasts longer, you have to remove the wheels and take out the axle bolt, and cover the bolt and wheel hubs with grease. Use the grease that auto service stations sell for the wheel bearings of an automobile. You'll only need to grease the wheels once every spring unless you use the machine in very dusty areas.

A light film of oil on the outside of the throttle cable—and any other enclosed cable on a machine—keeps it lubricated inside and working freely.

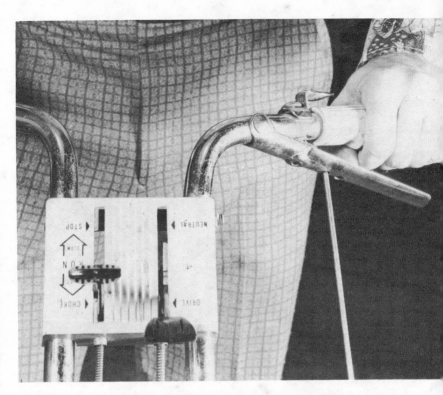

Here's an operating hint. Whenever you use a lawnmower or any machine, keep the throttle just below full speed. A lawnmower can cut any lawn at three-quarters throttle. Actually, most mowers do a neater job at lesser speeds.

The engine uses about half as much gasoline per hour as when you run the engine at full throttle. The savings over a summer's use are worth considering. Engine wear is cut down fantastically when you don't run the engine wide open all of the time.

If your mower cuts the lawn smoothly at half-throttle, then that's the speed to operate at. Whenever a bit of extra power is needed, say for heavier grass, the governor automatically opens the throttle enough to get through. Then it slows the engine back down to half-speed.

Any time you operate a garden tractor, lawnmower, or any other small-engine-powered machine, take some thought for safety. Keep your feet and hands clear of mower blades. Keep a lookout for children; they have a way of appearing suddenly from nowhere. Never allow anyone to stand on the grass outlet side of a lawnmower; any hard object thrown by a mower blade can be deadly. Keep garden tractors away from rough ground, you can be tossed from the machine very easily.

To sum it all up: take care of your small engines and machines, and . . .

DRIVE CAREFULLY

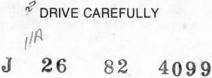